D0104654

HOW TO RENT A FIRE LOOKOUT IN THE PACIFIC NORTHWEST

A guide to renting fire lookouts, guard stations, ranger cabins, warming shelters and bunkhouses in the National Forests of OREGON and WASHINGTON

TOM FOLEY
TISH STEINFELD

WILDERNESS PRESS
BERKELEY

First Printing May 1996

Photos and maps by the authors except as noted
Design by Margaret Copeland
Cover design by Larry Van Dyke
Front cover photo © 1996 by David Steinfeld
Back cover photos © 1996 by Tom Foley

Library of Congress Card Number 96-7658
ISBN 0-89997-195-4

Manufactured in the United States of America

Published by Wilderness Press
2440 Bancroft Way
Berkeley, CA 94704
(800) 443-7227
FAX (510) 548-1355

Write, call or fax us for a free catalog

Front cover and frontispiece photos: Acker Rock Lookout, Umpqua National Forest, Oregon (see page 67).

Back cover photos: Indian Ridge Lookout, Willamette National Forest, Oregon (*top*, see page 57); Lost Lake Cabins, Mt. Hood National Forest, Oregon (*bottom*, see page 47).

♻ Printed on recycled paper with 20% post-consumer waste content

Library of Congress Cataloging-in-Publication Data

Foley, Tom, 1947-
How to rent a fire lookout in the Pacific Northwest / Tom Foley and Tish Steinfeld. — 1st ed.
p. cm.
ISBN 0-89997-195-4
1. Huts—Northwest, Pacific—Guidebooks. 2. Forest reserves--Northwest, Pacific—Guidebooks. I. Steinfeld, Tish, 1955- . II. Title.
TX907.3.P323F65 1996
647.94795—dc20 96-7658
CIP

Walking

It is true, we are but faint-hearted crusaders, even the walkers, nowadays, who undertake no persevering, never-ending enterprises. Our expeditions are but tours, and come round again at evening to the old hearthside from which we set out. Half the walk is but retracing our steps. We should go forth on the shortest walk, perchance, in the spirit of undying adventure, never to return, prepared to send back our embalmed hearts only as relics to our desolate kingdoms. If you are ready to leave father and mother, and brother and sister, and wife and child and friends, and never see them again—if you have paid your debts, and made your will, and settled all your affairs, and are a free man—then you are ready for a walk.

— Henry David Thoreau, *Walking*

Acknowledgements

A project such as this could not be completed without the help of a great many people—so many that to make a list of their names would take several pages. But we must thank Bruce Nichols of Bly Ranger District—without his help this book might not exist; Cherie Leonardo, also of Bly Ranger District; Catherine Callaghan of Lakeview Ranger District; Jackie McConnell of Bear Valley Ranger District; Brenda Taylor and Mel Ford of Barlow Ranger District; Susan Graham of Hood Canal Ranger District; Janel Lacy of Heppner Ranger District; Mike Keown of Illinois Valley Ranger District; and Harvey Timeus and Angie Dillingham of Chetco Ranger District.

Also, our heartfelt thanks to David Steinfeld, Veronica and Nino Foley, Pat McFadden, Gail Throop, Kevin Peer and Thomas Doty, who assisted in a multitude of ways—from the practical to the philosophical. Our special thanks to poet Gary Snyder, a former fire-lookout guard, who graciously allowed us to include his poems. Thanks also to Brian and Cathy Freeman, of Crystal Castle Graphics, for their generosity in sharing their expertise.

Our thanks to the U.S. Forest Service personnel who renovate and preserve these rustic structures and make them available for public use. And our thanks too, to the many community volunteers who donate time, materials, and support to this rare cause.

Tish Steinfeld
Tom Foley
May 1996

Table of Contents

Preface

It is surprising how many great men and women a small house will contain.
— Henry David Thoreau, *Walden*

This guide to small houses in the woods is, to the best of our knowledge, the only one of its kind. But there is a master guidebook, written by a master, on how to live in a small house in the woods. It was written in 1845, rewritten at least seven times thereafter, and was published, finally, in 1854. It is called *Walden* by Henry David Thoreau.

The thoughts expressed therein caused not a ripple on the surface of the waters of the times in which he lived, but later they influenced the thinking of millions—Martin Luther King and Mahatma Gandhi among them.

Thoreau, Ralph Waldo Emerson, Walt Whitman, and many of the thinkers of their era thought it was of the utmost importance occasionally to remove oneself from society and all its distractions, by repairing for however long was necessary or possible to a place that was secluded and relatively undisturbed by the works of humans—to the natural world. Here is how Thoreau put it: *"I went to the woods because I wished to live deliberately, to front only the essential facts of life, and see if I could not learn what it had to teach, and not, when I came to die, discover that I had not lived."*

The first wood in my own life was a small grove of old trees that lined the long avenue that led into our farmyard in a part of rural Ireland that was more or less treeless. As soon as I learned to walk I wandered—sauntered—daily, sometimes hourly, into that groved world. This habit of sauntering stayed with me and sustained me through my childhood and remained with me through my later life in Australia, South Africa, Botswana, England, Denmark, and, now, America.

In America I discovered that all the while I had, unwittingly, been following an old and exalted tradition—older than the great saunterers themselves—Thoreau, Emerson, Whitman. In

America, in the days before the dawn of that new species, Homo Television, the log cabin in the woods, and the way one chose to negotiate the trails and trials that led to it, epitomized all that was straight and true in the American heart. The log cabin is still there. The trails are still there. The woods are still there. This book's purpose is to lead you to them.

The trials, of course, are still with us also. But the hope is that, having found our cabin in the woods or our lookout on the mountaintop, we will allow ourselves to be taken where neither this book nor any other can take us, but where nature can. John Muir's assertion that "The clearest way into the Universe is through a forest wilderness" may then be profoundly meaningful to us.

We may begin to see that each one of us is as much a part of nature as the rock-wren, the spotted owl, or the Sierra Nevada, and that the sedulous appeals of organizations who dedicate themselves to protecting nature, based as they often are on negative emotions—fear, anger, guilt—are unlikely, in the long run, to do little other than produce more fear, anger, and guilt.

Wordsworth told us that "we come trailing clouds of glory." He did not exclude the owners and managers of logging companies. Dylan Thomas says that even in death, "we shall have stars at elbow and foot." If nature is whole and holy, so too are we. The whole of nature—from Mt. Whitney to Mt. Everest and all the oceans and rivers in between—it is all coursing in our veins, as close to us as the beating of our hearts.

We try to exile ourselves from it by looking at it from afar—on T.V. screens, in "nature" magazines, in "natural" history museums. We study it mercilessly under a microscope and give each other degrees in it. But the depth and breadth and scope and power of nature within each of us is ineffable. It is, at once, personal to each of us as well as universal to us all. It is the well from which we drink. Those who are willing to explore the depths of their own well may find that, though it is their own, its source is also the source of all wells: there is only one source.

It is our hope that a solitary sojourn in any of the cabins or fire lookouts presented in this book will allow the time and peace necessary to peer into the depths of that well.

Within it we may be able to see in a new light those we oppose so self-righteously. Their source is ours, ours is theirs. We may be able to see that they have hearts and souls remarkably like our own—flawed on one side, perfect on the other. Even more remarkable, some would say miraculous, are the changes that take place when we see those we oppose as they actually are—fellow pilgrims.

These changes spring from a mysterious and awesome source. But, despite our culture's cherished beliefs, we do not always know what is best for ourselves, our enemies, or the earth. And if we could see beyond our own dark horizon even for an instant, we might be unable to draw any line of separation between ourselves and our enemies, or the earth.

The spirit of the great forests of the Pacific Northwest was revered for thousands of years by generation after generation of Native Americans. By some estimates less than ten percent of these forests remains. But what remains of them and their healing spirit is still resplendent with redemption for us. And that redemption is offered freely, blindly, without reservation or qualification, and most certainly without accusation. The mountain streams speak incessantly of their eagerness to redeem us from the chaos and madness of our lives. The aspen groves still shimmer with the breath of the divine. The flower bows its head in scented hope.

And the womb of that hope is not science, but reverence.

We do not know a single blade of grass. A leaf confounds us. We know nothing, but that is all we need to know. We most certainly know not what we do. That is our saving grace. When we embrace it, there is not a pine needle in the forest that is not fragrant with forgiveness.

Tom Foley
Lookout Rock Lookout
Fremont National Forest
Oregon
1996

Quick Reference Chart

Rental	Geographic Region	National Forest	Ranger District	Phone Number	Season Available	Cost/ Night	Max Capacity	Page No.
1. Hamma Hamma Cabin	Western WA	Olympic	Hood Canal	360-877-5254	Year-round	$35	6	7
2. Interrorem Ranger Cabin	Western WA	Olympic	Hood Canal	360-877-5254	Year-round	$25	4	13
3. Burley Mountain Lookout	South-West WA	Gifford-Pinchot	Randle	360-497-1100	Nov-May	$20	4	18
4. Peterson Prairie Guard Station	South-West WA	Gifford-Pinchot	Mt. Adams	509-395-3400	Dec-May	$25+	6	23
5. Marble Mountain Warming Shelter	South-West WA	Gifford-Pinchot	St. Helens	206-750-3900	April-Oct	$40	100	27
6. Five Mile Butte Lookout	North-West OR	Mount Hood	Barlow	541-467-2291	Nov-May	$25	4	31
7. Flag Point Lookout	North-West OR	Mount Hood	Barlow	541-467-2291	Nov-May	$25	4	36
8. Valley View Cabin	North-West OR	Mount Hood	Barlow	541-467-2291	Nov-May	$25	4	42
9. Lost Lake Cabin	North-West OR	Mount Hood	Hood River	541-386-6366	May-Oct	$50	15	47
10. Warner Mountain Lookout	West-Central OR	Willamette	Rigdon	541-782-2266	Dec-May	$25	4	52
11. Indian Ridge Lookout	West-Central OR	Willamette	Blue River	541-822-3317	May-Oct	$20	4	57
12. Box Canyon Guard Station	West-Central OR	Willamette	Blue River	541-822-3317	May-Nov	$25	6	62
13. Acker Rock Lookout	South-West OR	Umpqua	Tiller	541-825-3201	June-Oct	$40	4	67
14. Pickett Butte Lookout	South-West OR	Umpqua	Tiller	541-825-3201	Nov-May	$40	4	72
15. Whisky Camp	South-West OR	Umpqua	Tiller	541-825-3201	June-Oct	$40	4	76
16. Pearsoll Peak Lookout	South-West OR	Siskiyou	Ill. Valley	541-592-2166	June-Oct	$20	4	80
17. Snow Camp Lookout	South-West OR	Siskiyou	Chetco	541-469-2196	May-Oct	$30	4	86
18. Packers Cabin	South-West OR	Siskiyou	Chetco	541-469-2196	Year-round	$20	12	92
19. Ludlum House	South-West OR	Siskiyou	Chetco	541-469-2196	Year-round	$20	15	97
20. Ditch Creek Guard Station	North-East OR	Umatilla	Heppner	541-676-9187	Year-round	$35	4	102
21. Clearwater Lookout Cabin	South-East WA	Umatilla	Pomeroy	509-843-1891	Year-round	$25	4	106

Rental	Geographic Region	National Forest	Ranger District	Phone Number	Season Available	Cost/ Night	Max Capacity	Page No.
22. Godman Guard Station	South-East WA	Umatilla	Pomeroy	509-843-1891	Year-round	$25+	10	110
23. Wenatchee Guard Station	South-East WA	Umatilla	Pomeroy	509-843-1891	Year-round	$25	4	114
24. Pearson Meadows Guard Station	North-East OR	Umatilla	John Day	541-427-3231	Year-round	$25	4	118
25. Fry Meadow Guard Station	North-East OR	Umatilla	Walla-Walla	509-522-6290	Year-round	$25	4	122
26. Summit Guard Station Bunkhouse	North-East OR	Umatilla	Walla Walla	509-522-6290	Year-round	$25	4	127
27. Two Color Guard Station	North-East OR	Wallowa-Whitman	La Grande	541-963-7186	Year-round	$40+	12	131
28. Antlers Guard Station	North-East OR	Wallowa-Whitman	Unity	541-446-3351	Year-round	$25	6	136
29. Peavy Cabin	North-East OR	Wallowa-Whitman	Baker	541-523-1932	Year-round	$40	4	140
30. Lily White Guard Station	North-East OR	Wallowa-Whitman	Pine	541-742-7511	Year-round	$40+	15	145
31. Murderer's Creek Guard Station	East-Central OR	Malheur	Bear Valley	541-575-2110	Nov-May	$25	4	149
32. Fall Mountain Lookout	East-Central OR	Malheur	Bear Valley	541-575-2110	Year-round	$25	2	153
33. Dry Soda Lookout	East-Central OR	Malheur	Bear Valley	541-575-2110	Nov-June	$25	2	157
34. Crane Prairie Guard Station	East-Central OR	Malheur	Prairie City	541-820-3311	Jan-April	$25	4	161
35. Short Creek Cabin	East-Central OR	Malheur	Prairie City	541-820-3311	Jan-April	$25	4	165
36. Bear Valley Work Center	East-Central OR	Malheur	Bear Valley	541-575-2110	Nov-May	$25	8	169
37. Flag Tail Lookout	East-Central OR	Malheur	Bear Valley	541-575-2110	Nov-June	$25	2	173
38. Deer Creek Guard Station	East-Central OR	Malheur	Bear Valley	541-575-2110	Nov-June	$25	4	177
39. Hager Mountain Lookout	South-Central OR	Fremont	Silver Lake	541-576-2107	Nov-May	$25	4	181
40. Bald Butte Lookout	South-Central OR	Fremont	Paisley	541-943-3114	Year-round	$25	4	186
41. Aspen Cabin	South-Central OR	Fremont	Lakeview	541-947-6359	June-Oct	$25	6	192
42. Fremont Point Cabin	South-Central OR	Fremont	Silver Lake	541-576-2107	Year-round	$25	4	196
43. Totem Bunkhouse	North-East OR	OR State Park	Emigrant Sp.	800-452-5687	Year-round	$15	4	201

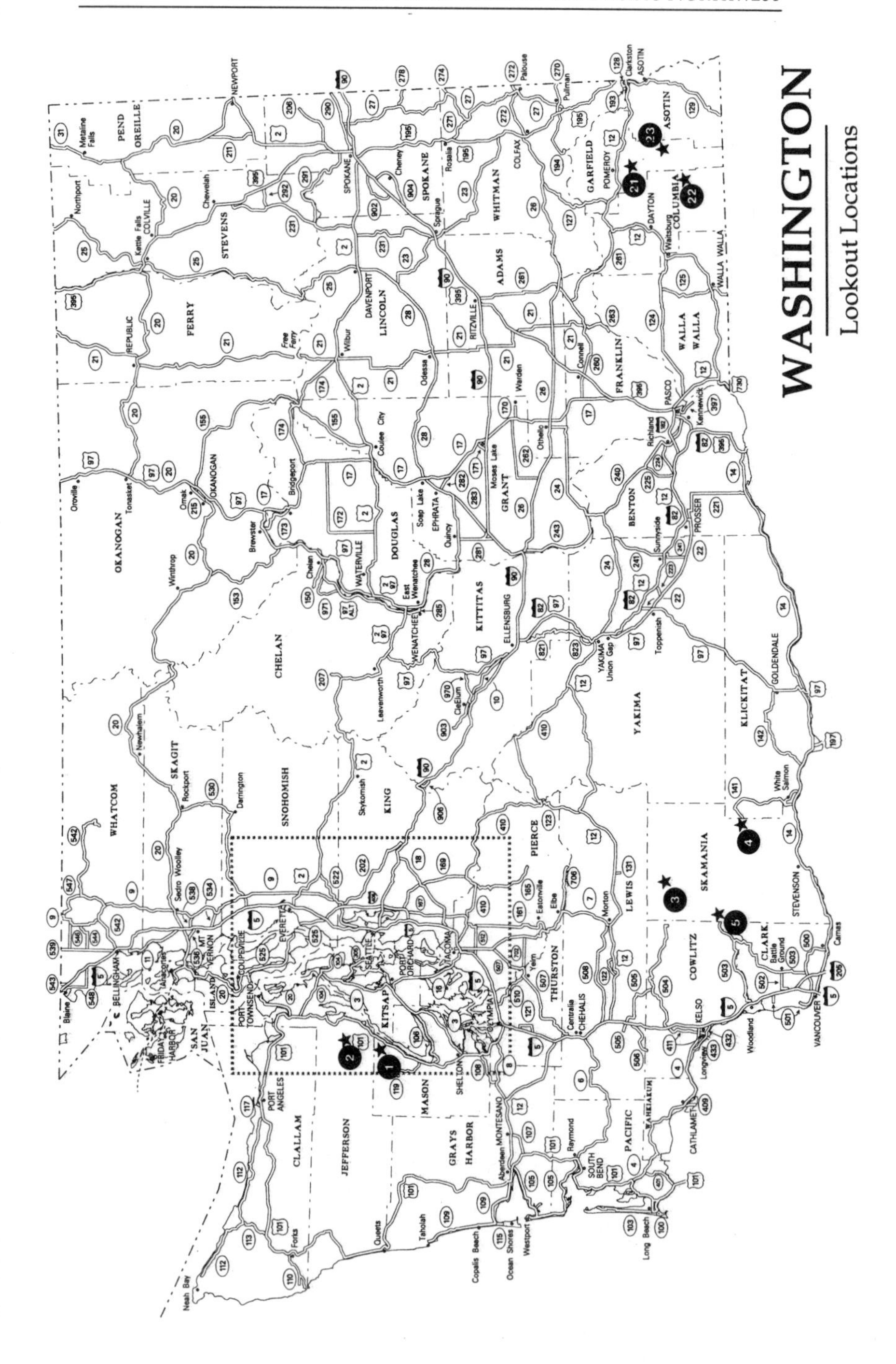
WASHINGTON
Lookout Locations
WHATCOM
OKANOGAN
FERRY
STEVENS
PEND OREILLE
SKAGIT
CHELAN
SNOHOMISH
KING
KITTITAS
DOUGLAS
LINCOLN
SPOKANE
GRANT
ADAMS
WHITMAN
GARFIELD
COLUMBIA
ASOTIN
FRANKLIN
WALLA WALLA
BENTON
YAKIMA
KLICKITAT
SKAMANIA
LEWIS
PIERCE
THURSTON
COWLITZ
CLARK
PACIFIC
WAHKIAKUM
GRAYS HARBOR
MASON
KITSAP
JEFFERSON
CLALLAM
ISLAND
SAN JUAN

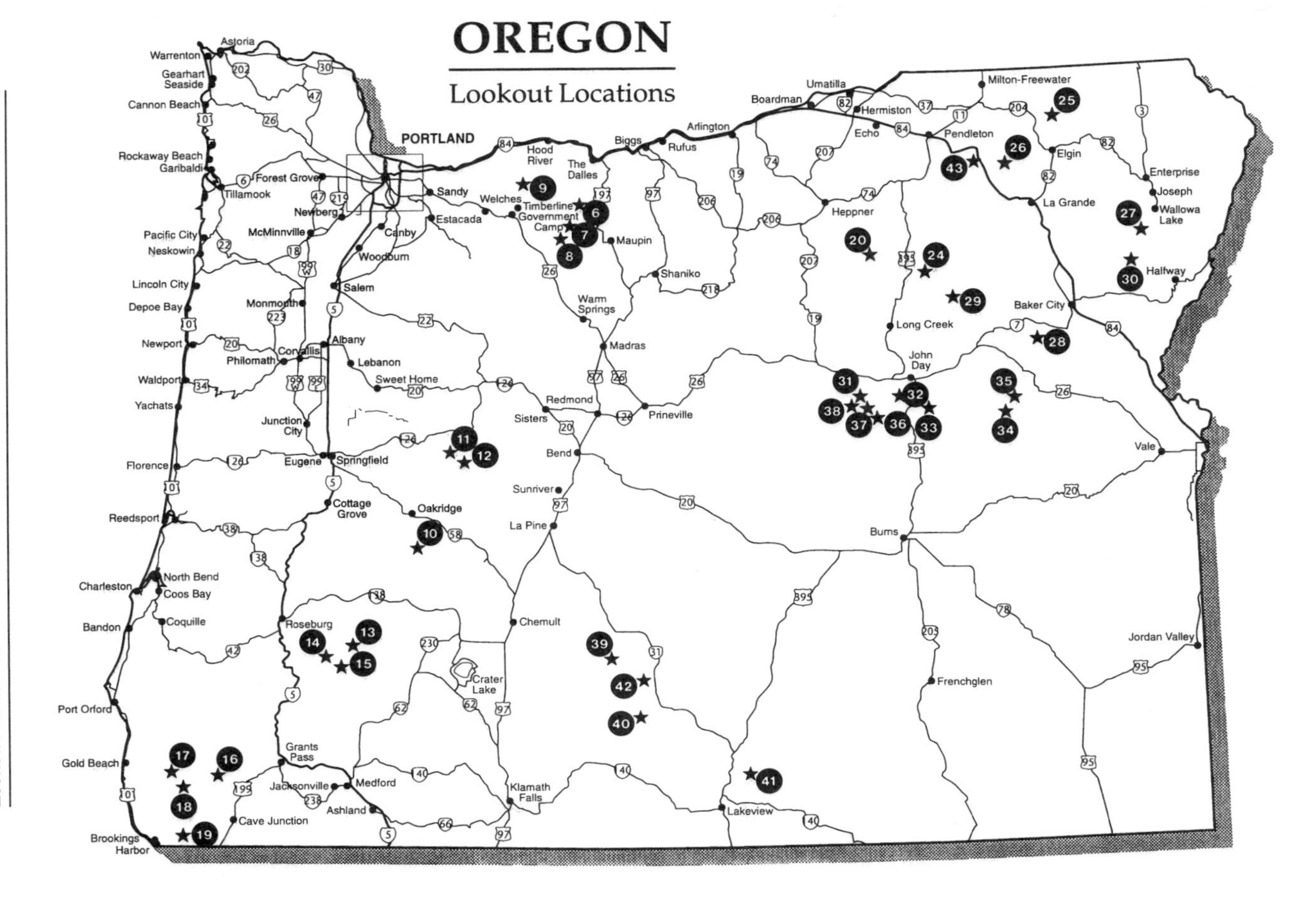
OREGON
Lookout Locations
PORTLAND
Warrenton
Astoria
Gearhart
Seaside
Cannon Beach
Rockaway Beach
Garibaldi
Tillamook
Forest Grove
Pacific City
Neskowin
McMinnville
Newberg
Lincoln City
Depoe Bay
Newport
Monmouth
Salem
Woodburn
Canby
Sandy
Estacada
Waldport
Yachats
Philomath
Corvallis
Albany
Lebanon
Sweet Home
Florence
Junction City
Eugene
Springfield
Reedsport
Cottage Grove
Oakridge
Charleston
North Bend
Coos Bay
Coquille
Bandon
Roseburg
Port Orford
Gold Beach
Brookings Harbor
Cave Junction
Grants Pass
Jacksonville
Medford
Ashland
Crater Lake
Klamath Falls
Chemult
La Pine
Sunriver
Bend
Sisters
Redmond
Prineville
Madras
Warm Springs
Welches
Timberline
Government Camp
Hood River
The Dalles
Maupin
Biggs
Rufus
Shaniko
Arlington
Boardman
Umatilla
Hermiston
Echo
Heppner
Pendleton
Milton-Freewater
Long Creek
John Day
Burns
Lakeview
Frenchglen
Baker City
La Grande
Elgin
Enterprise
Joseph
Wallowa Lake
Halfway
Vale
Jordan Valley

August on Sourdough, A visit from Dick Brewer

You hitched a thousand miles
north from San Francisco
Hiked up the mountainside a mile in the air

The little cabin—one room—
walled in glass
Meadows and snowfields, hundred of peaks

We lay in our sleeping bags
talking half the night;
Wind in the guy-cables summer mountain rain.

Next morning I went with you
as far as the cliffs,
Loaned you my poncho—the rain across the shale—

You down the snowfield
flapping in the wind
Waving a last goodbye half hidden in the clouds

To go on hitching
clear to New York;
Me back to my mountain and far, far, west.

— Gary Snyder, *Back Country*

Enjoying Your Adventure

Safety

We urge readers to use common sense when evaluating their hiking and skiing skills. If we say a hike is arduous, it may be impossible for some though easy for others. If we say the road is rough, leave the Maserati at home. Skiing to some of these places in a winter snowstorm might be thrilling for some but fatal for others. You will usually find at least one knowledgeable person at every Ranger District who will give you the information you need about weather, access, road conditions, snow depth, and difficulties you are likely to encounter at any particular time of year.

Always be well-prepared when entering the backcountry, especially in wintertime. Carry tire chains and a shovel. During a typical winter you may be traveling on skis or snowshoes for as much as ten miles or as little as a few hundred yards, depending on your particular destination. A longer trip can be extremely difficult in snow and could take an entire day. Plan accordingly and start early. Although there might be clear road access to the cabin or lookout on your way in, a heavy overnight snowfall could leave you stranded there, even in late fall or early spring. Be well-prepared. Carry extra clothing in waterproof containers. Be sure to notify someone of your destination, including departure and return dates.

If you intend bringing young children to a fire-lookout tower, please inquire from the managing Ranger District whether it is safe to do so. Some of the lookouts are ideal for children of any age, while others, because of their height, steep stairways, and precipitous, rocky surroundings are not.

Occasionally, during strong winds, lookout towers may sway slightly. Don't worry, they are built to do this. It is safer to remain in the tower than to attempt to descend the stairway during lightning or a wind storm. The lookout is well grounded . . . you may not be.

Many of the lookouts and cabins are equipped with propane appliances—heating, cooking, lighting. Use caution when using these and please remember to turn them off completely when leaving.

Water

Most lookouts and cabins do not have safe drinking water, and many have no water at all. Cleaning and washing water can sometimes be obtained from streams or melting snow; however, safe drinking water cannot be assured unless it is purified, filtered or boiled for five minutes. If you are bringing water, we suggest a gallon per day per person.

Lookouts

At most lookouts you will find the Osborne Fire Finder, a standard piece of equipment for observing and mapping smoke and wildfire. You are welcome to use it—with great care—for identifying landmarks and natural features of the area.

Rental Conditions

The Forest Service is eager to point out that it is not interested in the motel business. You will find no bed linen at any of these rentals—you may not find mattresses in some places—you will not find towels or chocolate mints on your pillow, in fact, you will not find any pillow. In most places there is no potable water—until you purify it—often no sink, and only in a few places will you find electricity or plumbing.

Reservations

First consult our *Quick Reference Chart,* and when you have decided on your destination, contact the managing Ranger District for an application packet, maps, and further information. The rentals are available on a first-come, first-served basis. Some Ranger Districts have specific check-in/check-out times, some do not. We specify those that do. Some Ranger Districts limit the duration of your stay to three nights, others allow as many as ten, others are undecided. We specify this information wherever possible. Again, check with the managing Ranger District.

For some rentals, such as Hamma Hamma, Interrorem, Two Color, Lost Lake, and Snow Camp, you may need to make reservations up to six months in advance, particularly on holiday weekends; however, advanced reservations are required for all rentals.

The rental fee for each structure is used exclusively for the maintenance of that particular structure, and is not lost, as you might expect, in the black hole of bureaucracy. Some Ranger Districts require a deposit, others do not. We specify those that do. If your plans change you may request a rental refund or a credit transfer, but you must do so at least three weeks before your reservation date. If the Forest Service determines the weather conditions are too severe for you to reach the rental safely, you will be given credit toward the next available date.

Remember to Bring

See the *Must Have* list below. Most cabins and lookouts are well supplied with a heat source, pots and pans, and eating utensils, though a few are not. Inquire at the managing Ranger District office. You will need the items listed in the *Must Have* section at all rentals. Our *What To Bring* heading for each structure includes only those items needed for that particular structure—though we always mention water where it is necessary.

Responsible Use

Please leave the cabin or lookout as you found it, or as you would like to have found it. Leave all pots, pans and utensils clean and ready for the next guests. Sweep the floor, pack out garbage, turn off all appliances, lights and propane, and lock the door.

If you are renting a lookout in the winter please leave the catwalk and steps clear of snow.

If you bring pets, make sure they don't disturb wildlife, and please clean up after them. If you bring stock animals, keep them tethered when not in use, and bring weed-free feed for them.

If you bring stoves or lanterns, we suggest propane fuel since it burns cleaner than liquid gas.

Most of these lookouts and cabins already have an established firering. Please use it; do not establish another. Campfires are a luxury. Please be kind to the woods by keeping your fire very small.

Hunters, please hang your kill outdoors (on meat poles when provided); please, please do not butcher any part of an animal in the cabin; and please pack out every last bit, leaving the area just as you found it.

Finally, no shooting within 1/4 mile of these rentals.

Note

Though we traveled thousands of miles visiting these destinations in an attempt to give you the best possible directions, due to ever-changing road names and numbers on Forest Service lands in the Pacific Northwest, and seasonal road closures, we urge readers to consult knowledgeable Ranger District personnel prior to each trip in order to get the most current information. We would greatly appreciate being notified, via Wilderness Press, of any changes that may have occurred since the book's publication. Your correspondence will help us improve later editions.

What To Bring

We always forget something when we plan for a trip—or we bring too much—but that is part of the experience of traveling. This list may help you decide what to bring and what not to bring. Some of these items are supplied at some rentals—check in the *What Is Provided* section of the cabin or lookout you are interested in renting.

Must Have

Special Use Permit
Drinking water and/or water filter
Food
Toilet paper
Waterproof matches
Garbage bags
Sunglasses
Pocket knife
Rental key
Sleeping bags
Ax and shovel
First-aid kit
Sleeping pads
Full tank gas
This guidebook
Flashlight
Good tires
Area/topographic maps
Extra batteries & bulb
Spare tire
Sunburn protection
Clothing (dress for extremes)
Change tire kit
Hat
Whistle
Compass

Nice To Have

Bug repellent
Ground coffee or tea bags
Bar soap
Day pack
Extra blankets
Dish towel
Salt and pepper
Pillows
Dish washing soap
Picnic cooler
Small gas stove (some places a must)
Water bucket
Eating utensils
Towels
Dishes, pots and pans (some places a must)
Lantern (some places a must)
Rope
Candles
Backpack

Useful Extras

Bikes
Pens and pencils
Nature guides
Binoculars
Writing pad
Star guide
Camera
Sketch pad, watercolors
Books
Musical instrument
Playing cards

Burning

Sourdough mountain called a fire in;
Up Thunder Creek, high on a ridge,
Hiked eighteen hours, finally found
A snag and a hundred feet around on fire:
All afternoon and into night
Digging the fire line
Falling the burning snag
It fanned sparks down like shooting stars
Over the dry woods, starting spot-fires
Flaring in wind up Skagit valley
From the Sound.
Toward morning it rained.
We slept in mud and ashes,
Woke at dawn, the fire was out,
The sky was clear, we saw
The last glimmer of the morning star.

— Gary Snyder, *Myths and Texts*

Fireguard, Hager Mountain Lookout.

1 Hamma Hamma Cabin
Olympic National Forest

"I feel that as long as the Earth can make a spring every year, I can. As long as the Earth can flower and produce nurturing fruit, I can, because I'm the Earth. I won't give up until the Earth gives up."

— Alice Walker

Your Bearings

50 miles northwest of Olympia
70 miles west of Seattle, approximately (via ferry)
75 miles northwest of Tacoma
170 miles north of Portland

Availability Year-round, weather permitting.

Capacity Six people. No pets.

Description

Single-story cabin with gabled and hipped roof lines. Living room, kitchen, two bedrooms, full bath. A delightful and very popular lodge in a beautiful setting.

Cost

An almost incredibly low $35 per night for this mansion in the mountains, plus $25 refundable deposit.

Reservations

Permits are available for as many as seven nights. For an application and list of the cabin's available dates contact Hood Canal Ranger District. The reservation procedure may appeal to the gambler in you. A weekly drawing is held, and available dates are filled from the applications received—which must be in writing. If your date preference is drawn, you will receive a permit form in the mail. You will have 14 days to return the completed permit along with full payment for your intended stay.

If an opening is *not* available for any of your date preferences, you will receive a new application and a list of available dates. For an application packet, maps, and further information contact:

Hood Canal Ranger District
PO Box 68
Hoodsport, WA 98548-0068
360-877-5254

How To Get There

The road is paved all the way. During the winter months access may be limited to cross-country skis and snowshoes the final four miles, though Hood River Ranger District tells us this happens only rarely. Consult them regarding current road and snow conditions prior to your departure.

From Hoodsport, travel 15 miles north on US Highway 101 to Forest Road 25. Turn left. The sign reads HAMMA HAMMA RECREATION AREA. Continue on the Hamma Hamma Road six miles to a driveway on the right. Watch for the sign on the right: HAMMA HAMMA CABIN: OCCUPIED RESIDENCE. The cabin is about 100 yards up this driveway.

Elevation 560 feet.

Map Location Olympic National Forest, Township 24 North, Range 3 West, Section 7.

What is Provided

Hamma Hamma is the king of cabins. Here, you will be treated to a living room, kitchen, two bedrooms, and a full bath with shower and flush toilet.

It is comfortably furnished with couches, an arm chair, coffee table and a fine gas stove. The main bedroom has a double bed and a high ceiling with exposed rafters and knotty pine walls. The other bedroom has a bunk bed, which sleeps two, and a single bed.

The kitchen has a wooden drop-leaf table with a built-in seating bench, two sinks with faucets displaying signs to tell you that the water from them is not potable. Neither, for that matter, is its flow reliable. So you may or may not have running water. Either way, it's not for drinking. During the summer camping season potable water is available from a hand-pump well at Hamma Hamma Campground, a quarter of a mile west on Forest Service Road 25. Off-season, make sure to bring enough of your own to last your entire stay, or have the means to treat the available nonpotable water supply.

There are wooden floors, window shutters, dark wall paneling and propane lighting throughout the cabin, which give a rustic, almost Old World atmosphere. The space heater, cook range, refrigerator, lights, and hot-water heater are all propane-powered, the cost of which is included in your fee. An empty storage room beside the kitchen opens onto a solidly built covered walkway connecting the house to the garage and the woodshed. Outside, at the rear of the building is a lovely picnic table as well as a firering and a vault toilet.

There is plenty of parking space—probably enough for 20 cars—surrounded by moss-covered stone walls. In the parking area there is a stone map kiosk which, though no longer in use, is reminiscent of the glory days of the great Northwest forests.

What To Bring

Drinking water, or the means to treat the local water.

The Setting

If a contest were held to decide the most sought-after rental cabin in this book, Hamma Hamma would be the certain winner, and the runner up would surely be its sister cabin on the Olympic Peninsula, Interrorem.

There are compelling reasons for this popularity, beyond the proximity to Seattle. Hamma Hamma is a rare and delightful place where one easily feels at home and at peace. It is more akin to a lodge than a cabin; its lovely living room is embraced by a semicircle of bay windows overlooking the Hamma Hamma River drainage.

History

The skill and craftsmanship of the Civilian Conservation Corps, which constructed this fine Guard Station during 1936 and 1937, have earned Hamma Hamma Cabin a nomination to the National Register of Historic Places.

We were intrigued by the origin and meaning of the name "Hamma Hamma." Having dismissed our suggestion that it was of porcine origin, possibly the name of a Native pork pie, the ever-resourceful Susie Graham of the Hood Canal Ranger District told us that, originally, it was thought to be the Twana Indian name for "Stinky Stinky," but that further research indicates that it may be the Twana Indian name of the root of a rush that grows in the area.

Local Girl Scouts have undertaken the maintenance of the cabin since March 1992. Please help them by keeping the site as you found it—or, at least, as you would like to have found it.

Around You

The Olympic Peninsula and Hood Canal. To the west is Mt. Skokomish Wilderness; to the northwest, Brothers Wilderness; to the east, Hood Canal. Two miles west on Road 25 is the trailhead for Lena Lake, Trail 810. There is parking, a vault toilet and well water at Lena Creek Campground. It is less than three miles to the lake—and the junction with Trail 811, which takes you to Upper Lena Lake—and 3-1/2 miles to the junction with Brother Trail 821, which takes you several miles into Brothers Wilderness.

To reach Skokomish Wilderness, continue west from Hamma Hamma Cabin on Road 25 for about six miles to access Putvin

Trail 813, classed as "most difficult." Or travel eight miles west to Trail 822, also classed as "most difficult," which leads to Mildred Lakes, 4-1/2 miles from Hoodsport.

"The entire earth is but a leaf."
— Henry David Thoreau, *Walden*

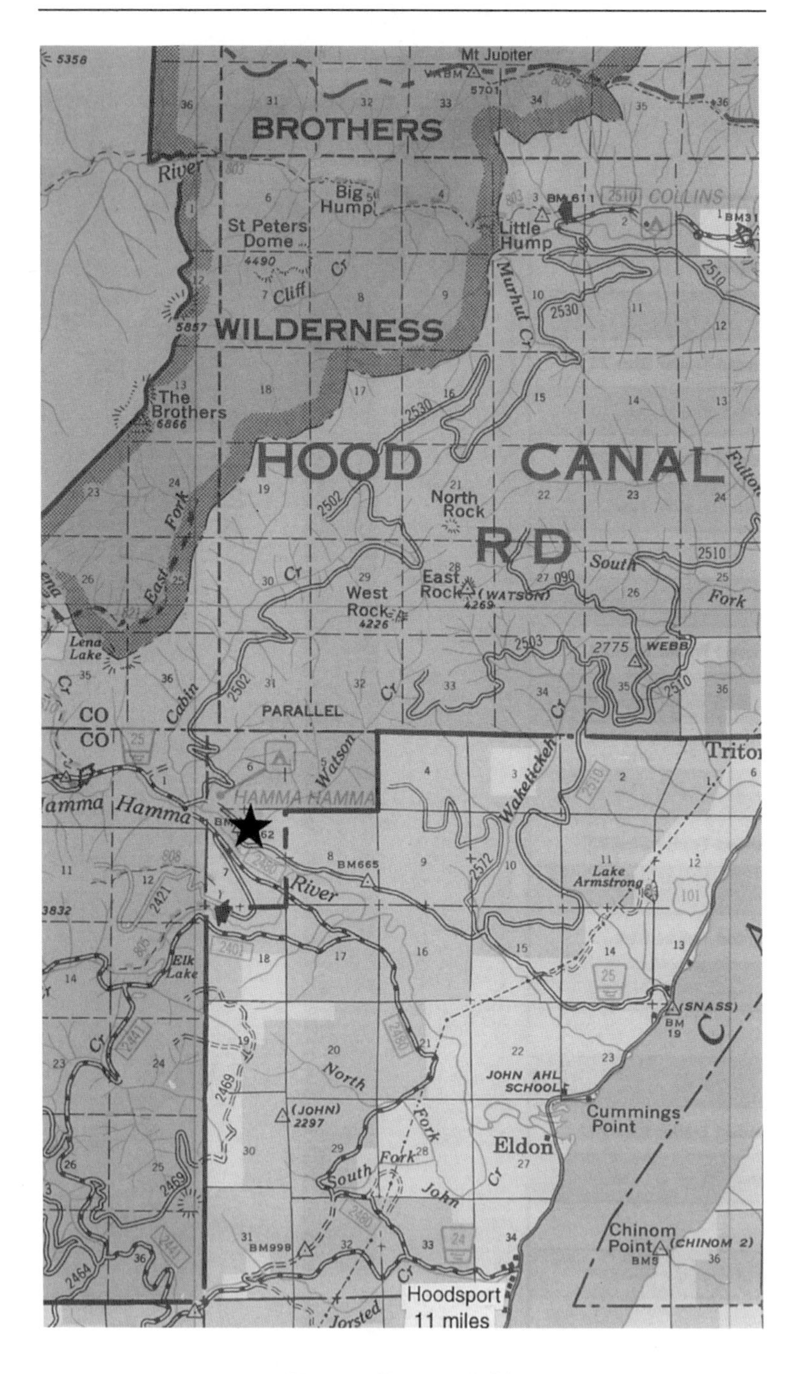

Hamma Hamma Cabin
Olympic National Forest, Washington

2 Interrorem Ranger Cabin

Olympic National Forest

"Now that I've refilled my heart and soul with the inspiration and tranquility of the forest, I can return to Seattle, normal."

— From the Cabin's Guestbook

Your Bearings

60 miles northwest of Olympia
75 miles west of Seattle
75 miles northwest of Tacoma
180 miles northwest of Portland

Availability Year-round.

Capacity A maximum of four people. No pets.

Description

Historic, single-story, 24x20-foot, peeled-log cabin with kitchen, living room, bedroom, open porch, and pyramidal cedar-shake roof. It is sturdy and honest, with an appealing gray hue, as though proud of its venerable age.

Cost $25 per night, plus a $25 refundable deposit.

Reservations

Permits are available for as many as seven nights. For an application and list of the cabin's available dates contact Hood Canal Ranger District. A weekly drawing is held, and available dates are filled from the applications received—which must be in writing. If your date preference is drawn, you will receive a permit form in the mail. You will have 14 days to return the completed permit and full payment for your intended stay.

If an opening is *not* available for any of your date preferences, you will receive a new application and a new list of available dates. For an application packet, maps, and further information contact:

Hood Canal Ranger District
PO Box 68
Hoodsport, WA 98548-0068
360-877-5254

This cabin's popularity is second only to that of its big sister, Hamma Hamma Cabin, which, though bigger, is considerably younger. Interrorem was built in 1907, Hamma Hamma in 1936.

How To Get There

Travel 23 miles north of Hoodsport on US Highway 101 to the Duckabush Recreation Area. Follow the Duckabush River Road, Road 2510, for four miles to the end of the pavement. Interrorem Ranger Cabin is very pleasantly situated in a fenced yard on the left side of the road.

Elevation 300 feet.

Map Location Olympic National Forest, Township 25 North, Range 3 West, Section 1.

What Is Provided

Propane fridge, cook range, heater and lights. Propane fuel is provided, water is not. During the summer camping season, potable water is available from a hand-pump well at Collins Campground, one mile west on Forest Road 2510. The kitchen has a vinyl floor, lots of closets, a stove, fridge, and two sinks—though no running water. The bedroom has two bunk beds; the living room has a painted wooden floor, a propane heater, exposed beams, and a table and chairs. Outside is a lovely yard, a picnic table, a vault toilet, a parking area, a barbecue and a fire-ring. And, if you're lucky, a herd of Roosevelt elk being chased by a friendly cat.

What To Bring

Drinking water and a bucket to bring water from Collins Campground, or have the means to treat river water.

History

Emery J. Finch, ranger and Hoodsport pioneer, built the cabin in 1907, and first occupied it with his new bride on April 22, 1908. It was the first administrative office site for Olympic National Forest.

Apart from its role as a honeymoon cabin, Interrorem has also hosted people in several government programs, such as the Emergency Relief Administration, the Works Progress Administration, and the Civilian Conservation Corps. From 1942 to 1986, it had yet another life as a fire-guard station, and from 1986 to 1994 it was used by Forest Service volunteers.

Through an extraordinarily felicitous coincidence the following entry appears in the guestbook to further the cabin's reputation as a honeymoon retreat:

> *We came here in January for a night and it was beautiful. This time we came back for the start of our honeymoon. We were married July 4th. [This is the day in 1845 that Henry David Thoreau moved into his cabin by Walden Pond. 'Yesterday I came here to live,' he wrote in his journal]. We stayed here July 4th and 5th and enjoyed a much needed rest . . . so thanks to the Forest Service for providing us with such a beautiful (and inexpensive) place to begin our married life, and we hope you all enjoy it as much as we have, and find it as healing.*

The origin and meaning of the name Interrorem are shrouded in the fog of the Olympic Peninsula. Despite its Latin sound, Susie Graham at the Hood Canal Ranger District suggested that it may not be Latin at all but, instead, may be a sly play on words. It seems that, originally, the cabin was built as an interim measure and that, somehow, the word interim became the much grander Interrorem.

Around You

The Olympic Peninsula, Hood Canal, Brothers Wilderness, and Olympic National Park.

From the Duckabush River Road there are lovely views of Brothers Wilderness and the Duckabush River, which may soon be designated a Wild and Scenic River. It is well known for its excellent fishing holes.

Ranger Hole Trail (824) takes off from beside the cabin and is an easy hike down to the Duckabush River, less than one mile away, where the rangers fished and got their household water. Be forewarned about the river, though. It is dangerous—replete with rapids and waterfalls.

The Interrorem Nature Trail (804) is a well-maintained 1/4-mile loop trail off of the Ranger Hole Trail (824). The vegetation here is similar to that found in the Olympic rain forest. There is an abundance of ferns and mosses as well as huge second-growth Douglas-fir and hemlock.

The Duckabush Trail (803) is reached by traveling another two miles west on Forest Road 2510—the Duckabush River Road in front of the cabin. This trail will take you into Brothers Wilderness—one mile from the trailhead and, if you keep going, into Olympic National Park in a distance of 6.8 miles from the trailhead. And if you still haven't had enough, it connects you with other trails within the Park.

As a World Heritage Park, Olympic National Park is in the exalted company of the pyramids of Egypt, Serengeti National Park of Tanzania and the Great Barrier Reef of Australia. It has the largest and finest example of virgin temperate rain forest in the Western Hemisphere and is one of the few coniferous rain forests in the world. It preserves the largest intact stand of coniferous forest in the contiguous 48 States, and a large herd of Roosevelt elk.

Detailed trail guides are available from the Hood Canal Ranger District. They will gladly mail one to you on request.

"The cost of a thing is the amount of what I call life which is required to be exchanged for it, immediately or in the long run."

— HENRY DAVID THOREAU

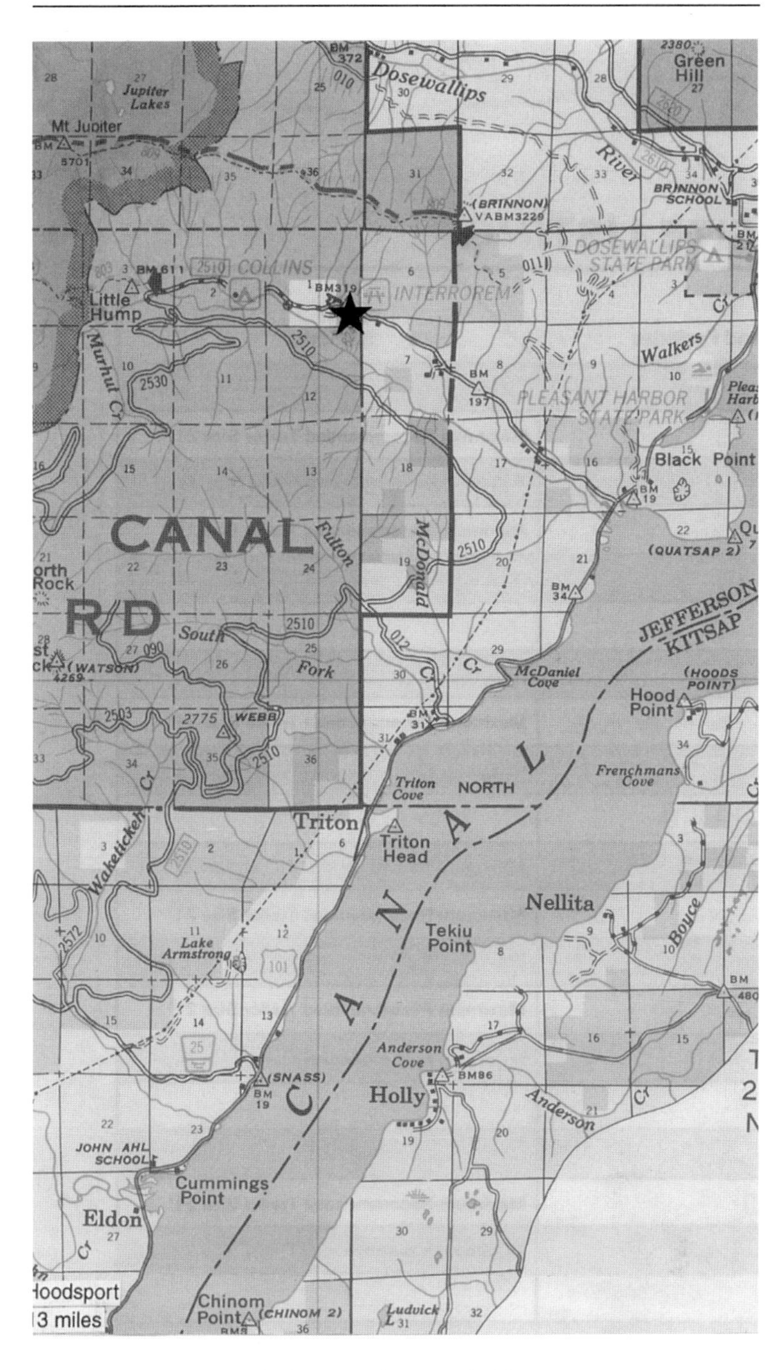

Interrorem Ranger Cabin
Olympic National Forest, Washington

3 Burley Mountain Lookout

Gifford Pinchot National Forest

"Beautiful morning, the sun is shining brightly and the clouds are all down low. We feel like we are in our own little kingdom above the clouds."

— FROM THE LOOKOUT'S GUESTBOOK

Your Bearings

30 miles south of Randle
90 miles southeast of Tacoma
115 miles southeast of Olympia
120 miles south of Seattle
155 miles northeast of Portland

Availability November 15 to May 30.

Capacity As many as four adults, but two would be more comfortable.

Description

14x14-foot, one-room fire lookout at ground level, with a hip roof. Known as an L-4, this type of lookout has heavy shutters hinged above the windows that can be propped open during summer for shade.

Cost $20 per night.

Reservations

Available for as many as seven consecutive nights. For an application packet, maps, and further information contact:

Randle Ranger District
10024 US Hwy. 12
Randle, WA 98377
360-497-1100

How To Get There

From Randle, Washington, take State Highway 131 south for about one mile. At the fork, veer left onto Forest Road 23 south, following signs for the Cispus Learning Center. This is Cispus Road. Proceed for about 11 miles to Forest Road 28. Turn right and go south 1-1/2 miles to Forest Road 76. Take another right here and remain on Road 76 for 3.5 miles to Forest Road 7605. Measure these 3.5 miles carefully, because Road 7605 is not signed —until you are actually on it. It is a most inconspicuous road.

And now that you have found Road 7605 you may want to stop for a minute while looking at the sign: HEAVY TRUCK TRAFFIC UNSUITED FOR PUBLIC TRAVEL, and ask yourself if you really do want to be on it. It is extremely narrow, extremely steep, extremely pot-holed and gutted in places, and, all that aside, one of the loveliest drives one can take. But when the Ranger District publications say it is "unsuitable for trailer traffic" they are showing a sly sense of humor. It is, in fact, unsuitable for anything but high-clearance vehicles—preferably with four-wheel-drive.

Should you decide not to travel this route, turn around and go back to Road 76, and travel west (left), for another mile until you reach Road 77. Turn left here on Road 77. This is a much better road than Road 7605, though it is gravel, steep and winding, and badly washboarded in places. It is also much longer.

Continue on Road 77 for 15 miles at which point you will turn left onto Road 7605. The advantage of choosing this route is that you will be on 7605 for only two miles. From 7605, turn right onto Road 086 (shown as Road 24 on the map) for another very steep and rocky mile. You will see the lookout straight ahead.

If you choose to try Forest Road 7605 from its junction with Road 76—it is seven miles less than via Road 77—follow its lovely, steep and languid way for 8.7 miles until you reach Road 086 (shown as Road 24 on the map). Turn left here. The lookout is one mile ahead.

If there is snow, expect a 3-7 mile ski trip to complete this route. Or, if following the Road 77 route, expect a 7-10 mile ski, depending on snow depth.

Elevation 5310 feet.

Map Location Gifford Pinchot National Forest, Township 11 North, Range 8 East, Sections 25 and 30.

What Is Provided

A propane cook stove, propane light, single bed with mattress, table and chairs, fire extinguisher, and map of the area. Staying here is a primitive experience in the wintertime: according to the Randle Ranger District the lookout does little more to protect you than slow down the wind. On the day we saw this lookout the traditional window shutters had been removed; in addition, Randle Ranger District advised us that for the foreseeable future, there would be no woodstove, no propane heater, and no fridge. The absence of a fridge will be the least of your worries. But the absence of a heat source and window shutters, in winter, at an elevation of 5310 feet, is not an experience we can recommend to anybody. But if you try it and survive, send us a postcard.

What To Bring

Drinking water is a must, or have the means to treat the local supply. Snow can be melted for your washing needs, but the Forest Service says that safe drinking water from snow cannot be assured. Extra food is a must; severe weather conditions may delay your intended departure.

Because of the possibility of fallen trees across the road, the Forest Service suggests that you bring a saw.

Prepare for harsh weather and pack accordingly. Because there is no heat source in the lookout, you will need to dress for extreme cold. And because of the possibility of the propane pipes freezing, you will need to bring your own camp stove and light source.

History

Constructed in 1934, it was once one of 60 fire lookouts in Gifford Pinchot National Forest. Today, it is one of only three that remain. The lookout was used as a fire lookout and radio relay center until 1972. It remained empty for the rest of the 1970s.

By the 1980s, this lookout was one of the few L-4 models that had not been burned down or dismantled. "I begged, pleaded, and cried big crocodile tears not to burn them down," said lookout guard Bud Panco. "They found it was cheaper to fly airplanes rather than pay for someone to be up at the lookout all season," said Walt Tokarczyk, another lookout guard.

Burley Mountain Lookout survived because of its central location on the Randle Ranger District, which gives it an important role in radio communication. For a number of years its primary purpose was to relay radio messages from one point to another; fire detection was, oddly, secondary.

The lookout was partially restored in the early 1980s, and today it serves two roles—fire detection in the summer and fall, and recreational retreat in the winter and spring.

Around You

Interpretive signs provide the names of the peaks as viewed from the lookout, including five Cascade volcanoes rising above the Cispus River Valley. To the south: Mt. Hood, Badger Peak, Pinto Rock, and Mount St. Helens. To the west: Riffle Lake, Kiona Rock, and Tower Rock. To the north: Mt. Rainier, Twin Sisters, and the Cispus River. To the east: Mt. Adams and Hat Rock.

The Burley Mountain Trail (256), three miles long, is a popular hike in the vicinity of the lookout. This trail crosses Forest Road 28 above the Cispus Learning Center and passes a waterfall at Covel Creek. It intersects Forest Road 7605, and from there continues another three miles on Road 7605 to the lookout. This is a fairly difficult hike with a very steep grade.

"Woke up at 6:18 am to the sunrise over a white sea of fluff with only bits of mountain tops sticking through—felt as though I was in another world."

— From the Lookout's Guestbook, May 9, 1994

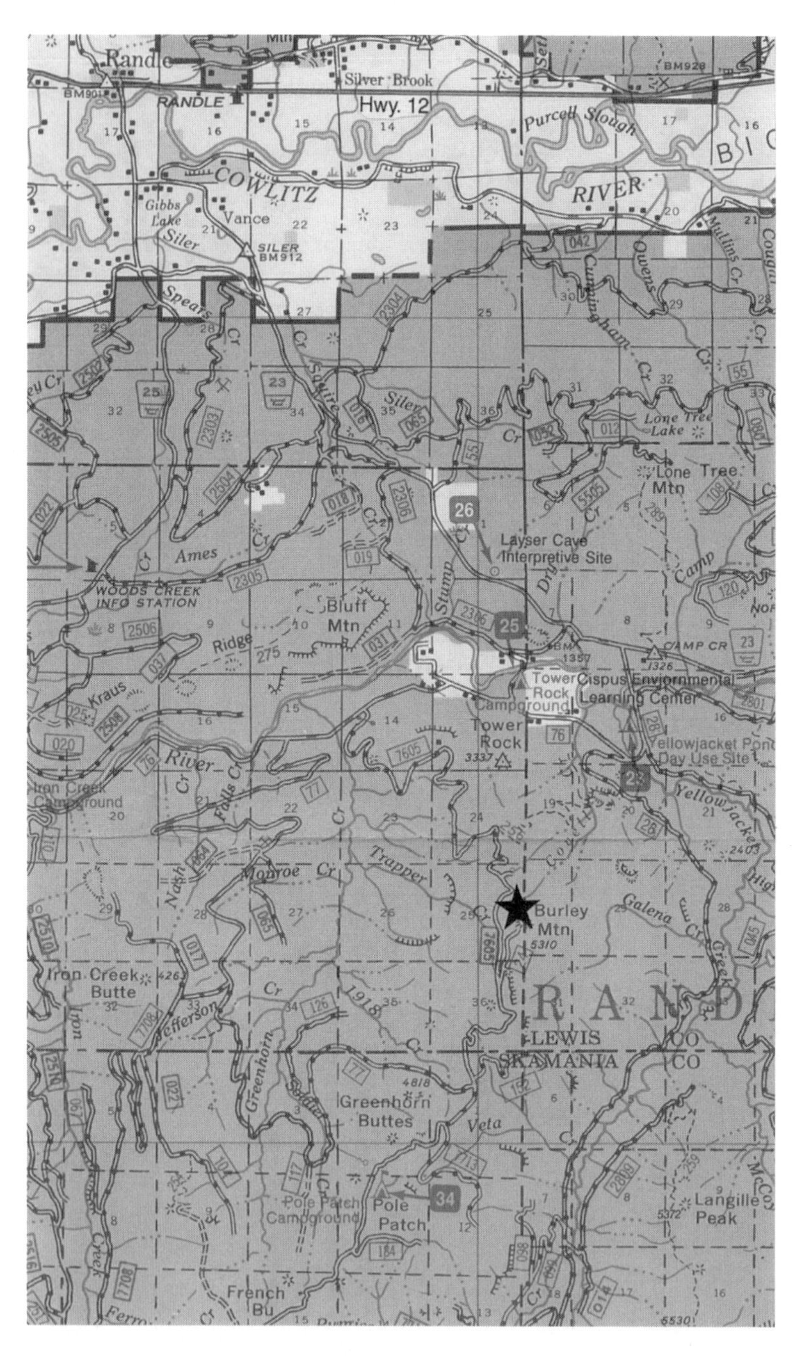

Burley Mountain Lookout
Gifford Pinchot National Forest, Washington

4 Peterson Prairie Guard Station

Gifford Pinchot National Forest

"As if you could kill time without injuring eternity."
— Henry David Thoreau, *Walden*

Your Bearings

10 miles southwest of the town of Trout Lake
35 miles north of the town of Hood River
60 miles northwest of The Dalles
95 miles northeast of Portland

Availability

December 5 to May 31.

Capacity The cabin has sleeping accommodations for four people, although a maximum of six is permitted.

Description

Delightful 18x24-foot cabin with bedroom, living room, and small kitchen, as well as an upstairs loft which, unfortunately, the Forest Service considers uninhabitable at present. There is a lovely old porch at the front, and a smaller one at the back.

Cost

$25 a night for two adults, with an extra charge of $5 per night for each additional adult. No charge for children under 12.

Reservations

Accepted from October 1 for as many as seven consecutive nights. For an application packet, maps, and further information contact:

Mount Adams Ranger District
2455 Highway 141
Trout Lake, WA 98650
509-395-3400

How To Get There

The cabin is accessible by vehicle on paved roads through late fall for as long as snow conditions allow, and again in the spring after the snow melts. During the snow season skis or snowshoes are required.

From Mount Adams Ranger District, just west of the town of Trout Lake, follow Highway 141 west and southwest for 6.7 miles to Atkisson Sno-Park. Here, at the boundary of Gifford Pinchot National Forest, the highway becomes Forest Road 24. The cabin is 2-1/2 miles further west on Forest Road 24, on the right side, about 1/4 mile past the main entrance to Peterson Prairie Campground.

Forest Road 24 is closed to wheeled vehicles on or about December 1, depending on snow accumulation. To leave your vehicle at Atkisson Sno-Park after November 15 requires a sno-park permit: a permit for one day costs $7, a three-day permit costs $10, and a season permit is $20. If you are able to drive all the way to the cabin, be aware that a heavy overnight snowfall could keep you stranded there.

Elevation 3000 feet.

Map Location Gifford Pinchot National Forest, Township 6 North, Range 9 East, Section 34.

What Is Provided

Propane cook stove, propane fuel and lanterns, bunk beds, woodstove, fire-extinguisher, firewood, ax, a map of the area, and a shovel.

The bedroom, about 8x8-feet, has bunk beds—sleeps two; the living room also has bunk beds—also sleeps two. It has a fine stone fireplace and chimney, two loveseats, two coffee tables and an arm chair. For lighting there are three propane wall lanterns.

The little kitchen has a woodstove, a 2-burner, counter-top propane stove, sink, and cabinets. The water is usually turned off during the winter months. There is a vault toilet at the back of the cabin.

What To Bring

Drinking water or the means to treat the local water. We recommend that you have an extra two to three days' supply of food and drinking water as weather may delay your intended departure.

History

This cabin was formerly a fire guard station, built in 1926 on the site of an even older log structure which was used by Forest Service rangers during backcountry patrols.

In the summer months, Peterson Prairie Guard Station is occupied by Paul and Marguerite Fleming, a couple in their eighties, who have been married for over fifty-five years. They volunteer their time with an organization called Recreation Unlimited. The few hours we spent with them is one of the treasures we carry with us from our long trek through Washington and Oregon ferreting out these cabins and fire lookouts.

Around You

This is a winter recreation area set aside for skiing and snowshoeing. About five miles north on Forest Road 24, near Little Goose Horse Camp, Trail 34 leads west through Indian Heaven Wilderness to the Pacific Crest National Scenic Trail in the heart of the Wilderness.

A trail guide is available at the Ranger District office that gives complete details on more than 50 other trails in the Mt. Adams Ranger District.

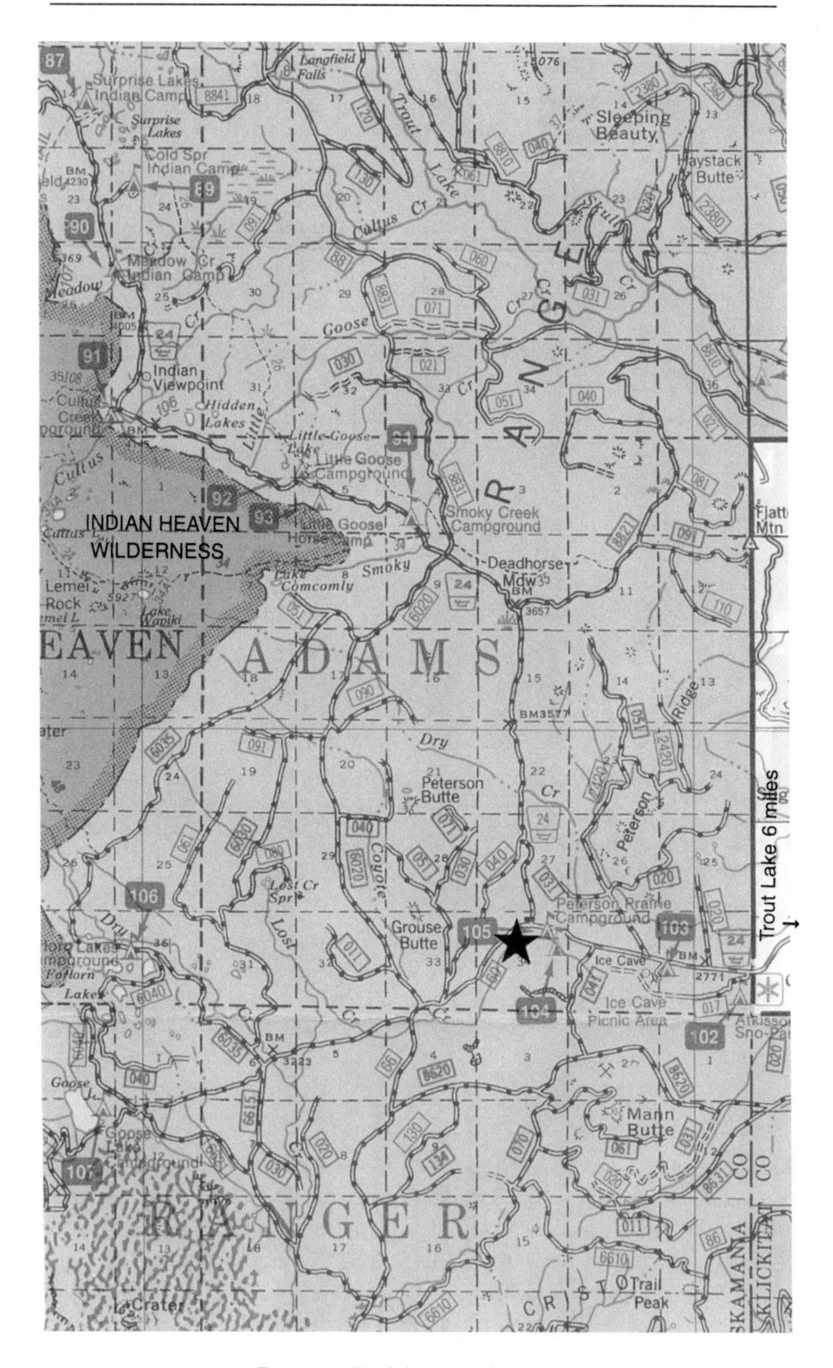

Peterson Prairie Guard Station
Gifford Pinchot National Forest, Washington

5 Marble Mountain Warming Shelter

Gifford Pinchot National Forest

"I know not the first letter of the alphabet. I have always been regretting that I was not as wise as the day I was born."

— Henry David Thoreau, *Walden*

Your Bearings

12 miles northeast of Cougar, WA
70 miles northeast of Portland
130 miles southeast of Olympia
160 miles south of Tacoma
190 miles south of Seattle

Availability

April 15 through November 1. For the remainder of the year it is available to skiers and snowshoers, rent-free, for day use.

Capacity

Shelters as many as 100 people. The paved parking lot accommodates 120 vehicles.

Description

30x30-foot log building with a metal roof, high ceiling, massive cross beams, poured-concrete floor, big double doors, and six plexiglass windows with folding shutters on each one: a discomforting alliance of antiquity and modernity. Both the shelter and restrooms are wheelchair-accessible.

Cost $40 per night.

Reservations For rental information contact:

Mount St. Helens National Volcanic Monument
42218 NE Yale Bridge Road
Amboy, WA 98601
206-750-3900

How To Get There

From Portland, travel 28 miles north on Interstate 5 to Woodland, Washington. Take Exit 21, and travel 28 miles east on State Highway 503, through the town of Cougar. (Highway 503 becomes Forest Road 90). From Cougar travel five miles east on Forest Road 90 to Forest Road 83. Turn left onto Road 83, following signs for Lahar Viewpoint and Lava Canyon Recreation Area. The Marble Mountain Sno-Park and Warming Shelter are on the left, six miles north on Road 83.

Elevation 2700 feet.

Map Location Gifford Pinchot National Forest, Township 8 North, Range 5 East, Section 34.

What Is Provided

Inside the structure there are five picnic tables and a woodstove. At the rear of the building is a covered porch where the wood is stacked. Though this structure is designed specifically as a warming shelter for winter recreation use, it is well suited to large gatherings during spring, summer and fall. The parking area accommodates motor homes, trailers and campers. There are two composting toilets close by that are wheelchair-accessible.

What To Bring

Drinking water or the means to treat local water, dishes, pots and pans, eating utensils, and lanterns. You may sleep in the shelter, but bunks are not provided.

History

Built in 1990 as the centerpiece of the Marble Mountain Sno-Park. Local snowmobile and cross-country ski clubs assist with yearly firewood cutting and site maintenance. Volunteer hosts are at the shelter on weekends during the winter recreation season.

Around You

There are dozens of trails in the vicinity, many of them leading into Mount St. Helens National Monument. A comprehensive trail guide is available at the Wind River Ranger District.

June Lake Trailhead (216B) is a mile or so east on Forest Road 83. Lahar Viewpoint and Lava Canyon Trailhead (184) are a few miles east of the shelter, on Forest Road 83. Both of these are wheelchair-accessible. Trails in the National Monument Area itself offer wonderful views of Mount St. Helens.

The world of course knows, but we will tell it yet again, that on Sunday morning, May 18, 1980, at 8:32, Mount St. Helens laid an egg in the shape of the north face of the mountain which slid and rolled in a rather ungainly fashion into Spirit Lake. Then, it was ill-mannered enough to cross a 1300-foot-high ridge, and rolled an additional 14 miles down the Toutle River.

Then its gaseous mother rather indelicately blew away 150 square miles of local forest, and belched up a mushroomed-shaped column of ash thousands of feet into the air—most inconsiderate of her and very unpleasant for the neighbors too.

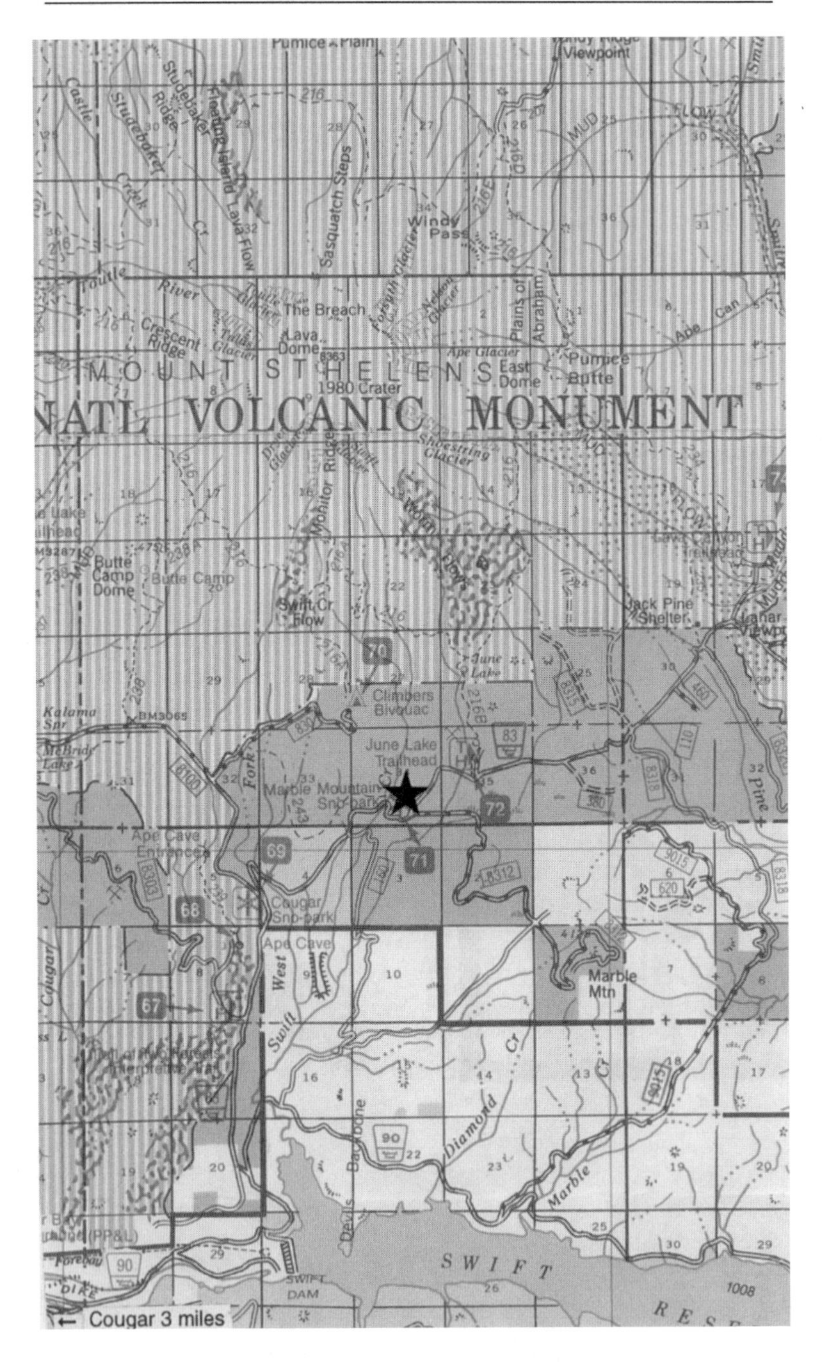

Marble Mountain Warming Shelter
Gifford Pinchot National Forest, Washington

6 Five Mile Butte Lookout

Mt. Hood National Forest

"I am not aware that we ever quarreled."

— Henry David Thoreau's response to a pious relative when asked, on his deathbed, if he had made his peace with God.

Your Bearings

20 miles west of Dufur
35 miles southeast of the town of Hood River
35 miles southwest of The Dalles
115 miles southeast of Portland via The Dalles and Dufur—though only 70 miles via Highway 26

Availability November 1 through May 31.

Capacity Four people maximum, though two would be more comfortable. Not suitable for small children.

Description

14x14-foot room with catwalk, atop a 30-foot tower. Storage shed with firewood and outhouse at ground level. Superb views.

Cost $25 per night.

Reservations

Available for as many as seven consecutive nights. For an application packet, maps, and further information contact:

Barlow Ranger District
PO Box 67
Dufur, OR 97021
541-467-2291

Safety Considerations

During a typical winter it will be necessary to leave your car in a parking area and travel by skis or snowshoes the final three miles to the lookout.

Access by skis takes approximately four hours. The route is marked with orange, arrow-shaped mileage signs. Be prepared for extreme weather conditions. Always contact the Ranger District prior to your departure for the latest road and weather conditions.

Because of the height of the lookout and its open catwalk, it may be risky to bring children. The stairway and catwalk are wooden and are quite slippery in rain, snow, and ice. Occasionally, during strong winds, the tower may sway slightly. It is built to do this. It is safer to remain in the lookout than to attempt to descend the stairway during wind storms or lightning. The lookout is well-grounded. Enjoy the spectacle.

How To Get There

Travel west on Dufur Valley Road from the town of Dufur for a little over 18 miles to Forest Road 4430, following the sign for RAMSEY HALL and CAMP BALDWIN. Dufur Valley Road becomes Dufur Mill Road, and then Forest Road 44. It is paved all the way.

You may park your vehicle here at the junction of Roads 4430 and 44. There are two routes from this point; one is three miles long, the other is four. The two are clearly marked, and together offer an excellent loop route to and from the lookout. (There is additional parking at Camp Baldwin, about a mile east).

To follow the first route, turn right on Forest Road 4430 for about 3/4 mile to Forest Road 120. Turn left onto Road 120, and after 1-1/2 to 2 miles, take another left onto Road 122, (not shown on map) from where you will see a green gate leading to the lookout.

The alternate route from the junction of Roads 44 and 4430 is less than four miles long, and takes you about two miles west on Road 44 (past its junction with Road 4430) to Road 120. Turn right onto Road 120 and follow this to Road 122 (not shown on map) and the green gate leading to the lookout. If you're lucky enough to be driving, neither Road 120 nor Road 122 is maintained—though they are negotiable, more or less.

Just north of Forest Road 44, on Forest Road 4430, there is a lovely campground set among tall firs, pines, and cedars stretched out languidly along Eight Mile Creek, called Eight Mile Crossing. Unfortunately, on the day we were there, there were trailers and mobile homes back to back and side to side, some of them nearly as big as an entire Irish Village.

Elevation 4627 feet.

Map Location Mt. Hood National Forest, Township 2 South, Range 11 East, Section 7.

What Is Provided

The lookout is immaculately clean and newly refurbished. There is one double bed, propane cook stove and fuel, table and chairs, fire extinguisher, woodstove, firewood, shovel, a few dishes and utensils, and maps of the area. A rope-and-pulley system transports gear up to the catwalk.

The 2-way radio is to be used for *emergency communications* only.

What To Bring

Drinking water is a must, or the means to treat local water. Snow can be melted for your washing needs, but the Forest Service says that safe drinking water from snow cannot be assured. Extra food is a must—severe weather conditions may delay your intended departure. Prepare for harsh weather and pack accordingly.

History

This lookout site was established around 1930. The present cabin was built in the early 1960s. This structure is still used as a fire lookout throughout the summer months.

Around You

Magnificent 360° views. To the north is snow-covered Mt. Adams and to the west, just 10 miles away, is Mt. Hood. A new trail now offers hikers access to the lookout from Eight Mile Creek.

Mid August at Sourdough Mountain Lookout

Down valley a smoke haze
Three days heat, after five days rain
Pitch glows on the fir-cones
Across rocks and meadows
Swarms of new flies.
I cannot remember things I once read
A few friends, but they are in cities.
Drinking cold snow-water from a tin cup
Looking down for miles
Through high still air.

Gary Snyder, *No Nature*

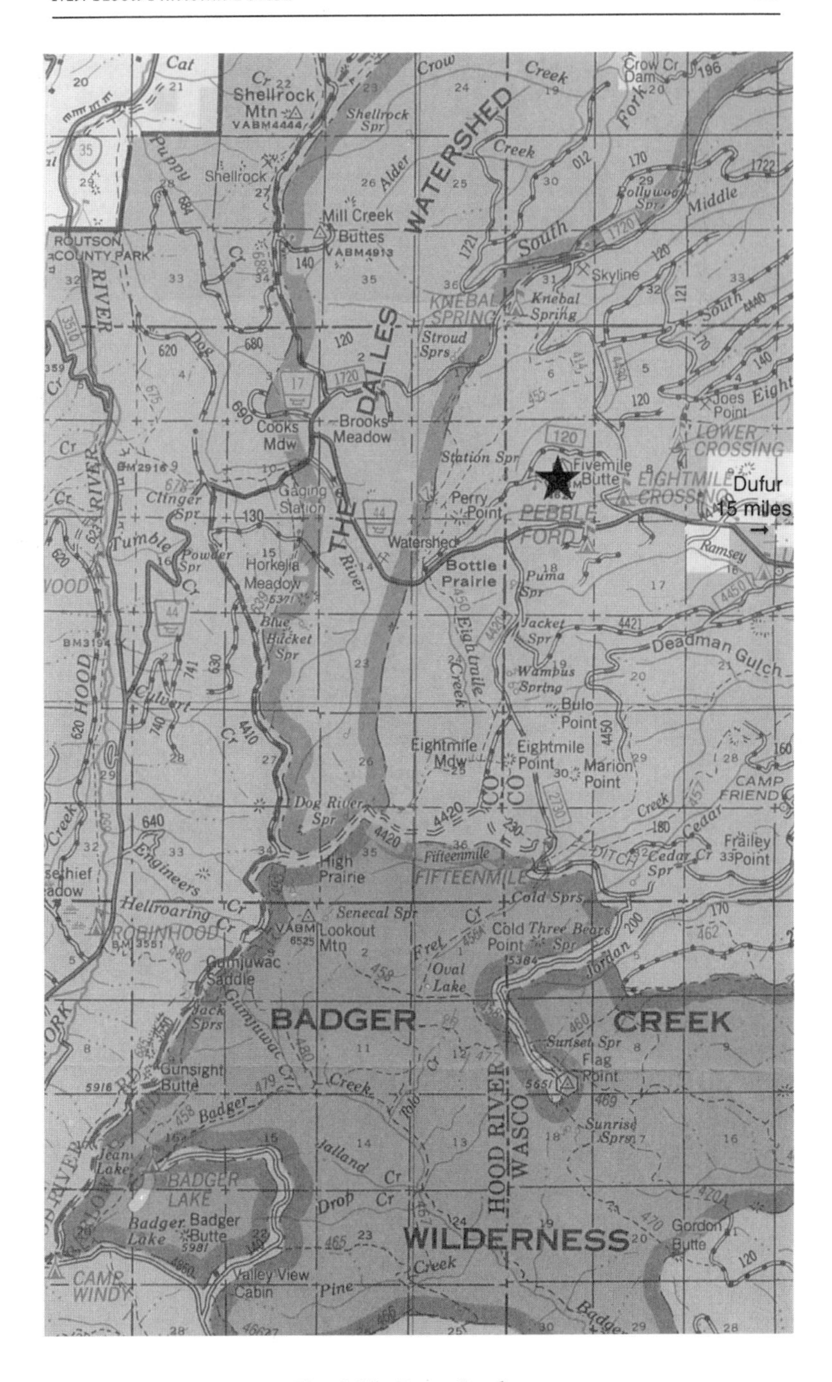

Five Mile Butte Lookout
Mt. Hood National Forest, Oregon

7 Flag Point Lookout

Mt. Hood National Forest

"Heaven parted and we floated gleefully until parting."

— FROM THE LOOKOUT'S GUESTBOOK

Your Bearings

30 miles southwest of Dufur
45 miles southwest of The Dalles
50 miles southeast of the town of Hood River
125 miles southeast of Portland via The Dalles and Dufur—though only 80 miles via Highway 26

Availability November 1 through May 31.

Capacity Four people maximum, though two would be more comfortable. Not suitable for small children.

Description

A clean and well maintained 14x14-foot room atop a 60-foot tower, with splendid views. A rope and pulley system transports gear up to the catwalk.

Cost $25 per night.

Reservations

Available for as many as seven consecutive nights. For an application packet, maps, and further information contact:

Barlow Ranger District
PO Box 67
Dufur, OR 97021
541-467-2291

Safety Considerations

During a typical winter it will be necessary to leave your car in a parking area and travel by skis or snowshoes the final 11 miles to the lookout. Access by skis takes a minimum of eight hours and is *not* a trip for novices.

The trail is marked with orange, arrow-shaped mileage signs. Be prepared for extreme weather conditions. You may need to spend a night out en route.

The height of the lookout—60 feet above the ground—as well as its open catwalk, make this rental risky for children, even with automobile access. Both stairway and catwalk are wooden and are slippery in rain, snow and ice. Occasionally, during strong winds, the tower will sway slightly. It's built to do this. It is safer to remain in the cabin than to attempt to descend the stairway during wind storms or lightning. The lookout is well grounded. Enjoy nature's fireworks.

How To Get There

Travel west on Dufur Valley Road from the town of Dufur for a little over 18 miles to Forest Road 4430, following the sign for

RAMSEY HALL and CAMP BALDWIN. Dufur Valley Road becomes Dufur Mill Road, and then Forest Road 44. It is paved all the way. The route is clearly signed. Winter access can range from moderate to difficult depending on snow conditions. You may park your vehicle at the junction of Roads 4430 and 44. (There is additional parking at Camp Baldwin, about a mile east).

Follow Forest Road 44 west to its junction with Forest Road 4420—about 1-1/2 miles. Turn left and follow Road 4420 for 2-1/2 miles—where it becomes Road 2730. Follow Road 2730 for 3-1/2 miles to Road 200: turn right here. The lookout is 3.7 miles ahead—though the sign says three miles.

Please consult the Ranger District regarding current road and snow conditions prior to your departure. All the roads are paved except for Road 200, which is not maintained—though it was navigable the day we drove it.

For those unwilling to go the whole way to the lookout in a single day there is a delectable little campground just beyond half way, at Fifteen Mile Creek, at the edge of Badger Creek Wilderness. There is space for only three tents here, one of which is on the banks of the creek. The creek is within the Dufur Municipal Watershed so please take good care of it. Fifteen Mile Trail (456) passes through here, and yes, it is 10.3 miles long.

Elevation 5650 feet.

Map Location Mt. Hood National Forest, Township 3 South, Range 11 East, Section 7.

What Is Provided

Propane cook stove and fuel, table and chairs, a single bed, fire extinguisher, firewood, shovel, a few dishes and utensils, and area maps. It has a small sink, a fridge with freezer, and a woodstove. The lookout is equipped with solar panels, and if the gods smile on you, you may have solar light for an hour or so, after dark of course.

Fifty-six steps take you, gasping, to the cabin, though a rope-and-pulley system will transport your gear straight up with blessed ease. An outhouse, and a storage shed containing firewood are on the ground below. The lookout is equipped with a satellite link telephone with instructions and emergency phone numbers posted nearby. This is to be used *only for emergency communications,* not for ordering pizza.

Like many of the Forest Service rentals, this lookout has a guestbook where visitors may leave something behind for those yet to come.

> *5-1-95 The sleep on the catwalk was great. Stayed toasty in the bag, and what a sunrise! Now it's back to civilization to pick up Dead tickets. Life is hell. P.S. Many thanks to the Forest Service for this great spot.*
>
> *3-27-94 Skied in by the light of a perfect full moon. Enjoyed two beautiful days of sunshine, blue skies, good friends and soft spring rain. Hard to imagine a cabin with a better view. Love this place.*
>
> *12-31-93 We celebrated New Year's Eve with champagne and OPB. Happy New Year, world. The trees danced in the wind outside, we danced inside. Glorious winter storm. Thank you Forest Service staff for making this available.*

What To Bring

Drinking water is a must, or the means to treat the local water. Snow can be melted for your washing needs, but the Forest Service says that safe drinking water from snow cannot be assured. Extra food is a must—severe weather conditions may delay your intended departure time.

Prepare for harsh weather and pack accordingly.

History

This lookout site was established in the 1930s and has been in continuous use since. The present cabin was built in the 1960s.

Around You

This fire lookout is unique in that it is the only one we have come across that is surrounded by wilderness. It is approximately 15 miles southeast of Mt. Hood in the North Cascade Mountains, and within Badger Creek Wilderness.

From the lookout you will see Mt. Hood, snow-covered of course, close up and to the west. Mt. Adams, also snow-covered, is to the north. On a clear day one can catch glimpses of Mount St. Helens, and, we're told, even Mt. Rainier. To the south: Mt. Jefferson and the Three Sisters.

Trails beside the lookout: Douglas Cabin Trail (470) takes you south to Sunrise Spring (.5 mile), and Helispot (2.2 miles). About a mile before reaching the lookout there are the following trails off Forest Road 200 which take you into the heart of Badger Creek Wilderness: Badger Creek Cutoff Trail (477), Divide Trail (458), and the Little Badger Trail (469). They are all are open to horses, for better or worse, and there is parking at the trailheads.

Mt. Hood from Flag Point Lookout.

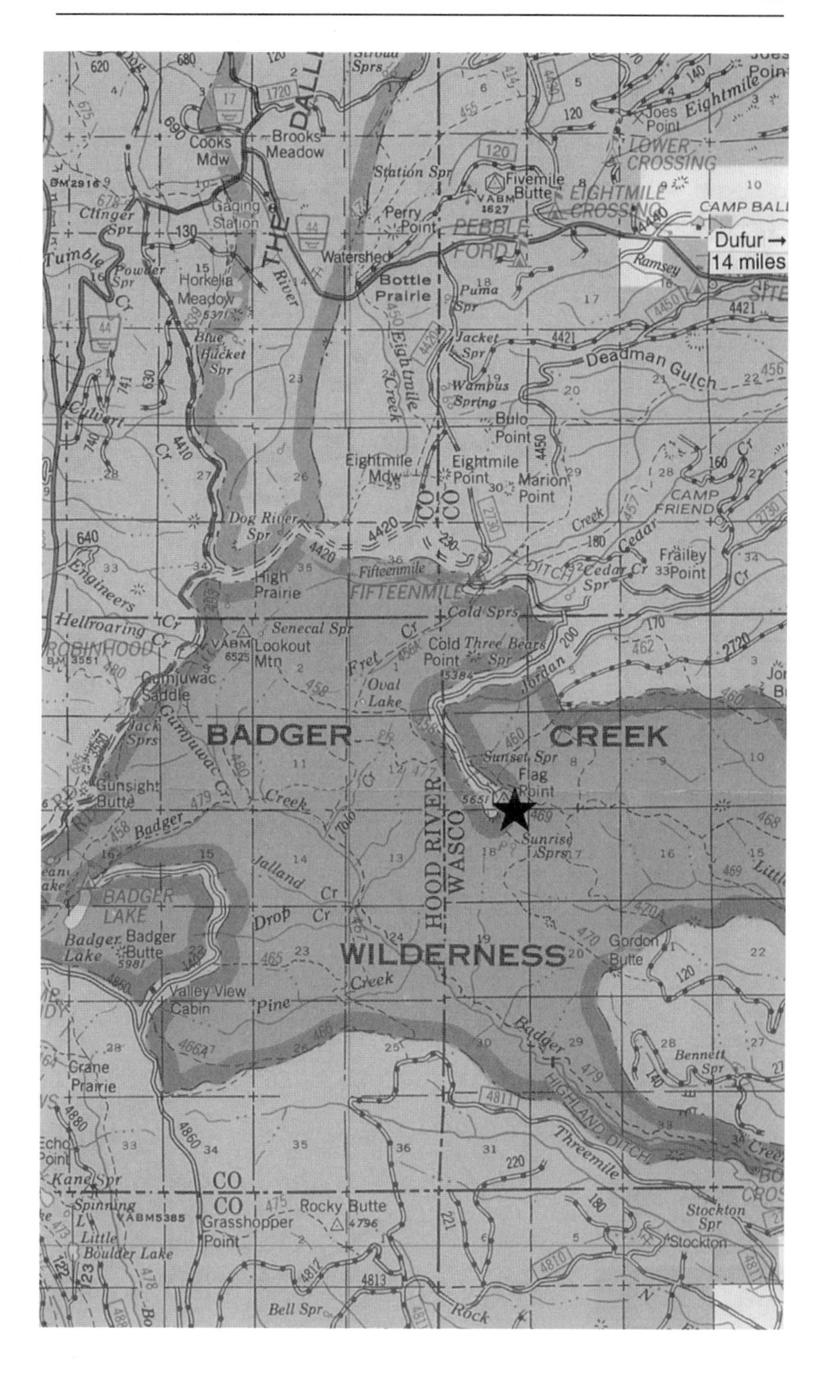

Flag Point Lookout
Mt. Hood National Forest, Oregon

8 Valley View Cabin

Mt. Hood National Forest

"I wish to hear the silence of the night, for the silence is something positive and to be heard."

— HENRY DAVID THOREAU, *Journal*, JANUARY 21, 1853

Your Bearings

40 miles southwest of Dufur
50 miles south of the town of Hood River
60 miles southwest of The Dalles
140 southeast of Portland via The Dalles and Dufur—though only 70 miles via Highway 26

Availability November 1 through May 31.

Capacity Four people maximum, though two would be more comfortable.

Description

One-room cabin measuring 12x18 feet, with a wooden floor. New metal roof, and new front and back porches. The interior looks somewhat distressed and in need of comforting. Your rental fee helps.

Cost $25 per night.

Reservations

Available for as many as seven consecutive nights. For an application packet, maps, and further information contact:

Barlow Ranger District
PO Box 67
Dufur, OR 97021
541-467-2291

Safety Considerations

During a typical winter you will be traveling by skis or snowshoes the last several miles to this winter retreat. Skiing time can vary from 4 to 8 hours, depending on snow levels and weather. The trip is suitable for moderately to highly experienced skiers, again depending on snow depths and weather.

Be prepared for extreme winter storms. Consult the Ranger District personnel immediately prior to your departure. Weather conditions can change quickly and dramatically.

How To Get There

From Dufur, Oregon, travel south on Highway 197 approximately 17 miles to the Tygh Valley exit and Tygh Valley Road, which becomes Wamic Market Road. Follow the signs for Tygh Valley and Wamic. After 6 miles, turn left at the town of Wamic, following the signs for ROCK CREEK RESERVOIR and BADGER LAKE, to Forest Road 48. Travel west on Road 48 for 10 miles, to Forest Road 4860. Turn right onto Road 4860.

At this point, there is a sign that reads: PRIMITIVE ROAD 3 MILES AHEAD. NOT RECOMMENDED FOR PASSENGER CARS OR TRAILERS. ALL TRAILERS PROHIBITED AT BADGER LAKE. Pay heed to this sign. This is a dreadful road—leave the Ferrari at home. After three miles the pavement ends and the gravel begins; two miles later the gravel ends and the dirt begins; and shortly after that it is for the birds—or at least for those with high-clearance 4x4 vehicles.

After eight miles on Road 4860 you'll be happy to veer right onto Spur Road 140. The cabin is one-half mile north from the fork, on the left side of the road.

On the way through the town of Wamic you may have noticed that, with commendable modesty, the good citizens make the claim that Wamic is THE GATEWAY TO BARLOW ROAD.

Your wheels are rolling through history as you drive west on what is now Forest Road 48. Some portions of this road were built in 1845-46 by one Samuel K. Barlow, from Kentucky. It was the first road over the North Cascades.

Elevation 5600 feet.

Map Location Mt. Hood National Forest, Township 3 South, Range 10 East, Section 21.

What Is Provided

A woodstove and firewood, two sets of bunk beds, a table and chairs, a light, a fire extinguisher, shovel, a limited quantity of dishes and utensils, coffee table, chest of drawers and closets, maps of the area, and a vault toilet.

What To Bring

Drinking water is a must, or the means to treat the local supply. Snow can be melted for your washing needs, but the Forest Service says that safe drinking water from snow cannot be assured. Extra food is a must: severe weather conditions may delay your departure. Prepare for harsh weather and pack accordingly.

Setting

The first thing to know about Valley View Cabin is that though there is a cabin and there is a valley, there is no view of the valley, or, indeed of anything else, for it is entirely surrounded by trees. But, if you've made it this far on skis, you are a hardy soul indeed—and will take with equanimity the honest simplicity of this place.

Valley View Cabin is unique in that it is the only ranger cabin we came across that is surrounded by a designated wilderness area—Badger Creek Wilderness, approximately 2-1/2 miles from Badger Lake.

History

Built in the early 1950s, and used by fire- suppression crews and wilderness guards during the summer months.

Around You

The cabin is seven miles southeast of Mt. Hood in the North Cascade Mountains, and within Badger Creek Wilderness. On the day we were there, in early July, this area was ablaze with wildflowers: lupine, camas, columbine, balsam root, paintbrush . . . whole hillsides of them.

Badger Lake is 2.5 miles northwest of the cabin on Road 140.

There are several hiking trails in the area that lead deeper into Badger Creek Wilderness. Pine Creek Trail (465), two miles long and less than a mile north of the cabin on Forest Road 14, connects to Trail 467.

Three Mile Trail (466) is, you guessed, four miles long, and a mile south of the cabin on Forest Road 4860. It connects to Trail 467. Mud Spring Trail (466A), 1-1/4 miles long and just south of the cabin on Forest Road 140, connects to Trail 466.

"Oh as I was young and easy in the mercy of his means,
Time held me green and dying though I sang in my chains like the sea."

— Dylan Thomas, *Fern Hill*

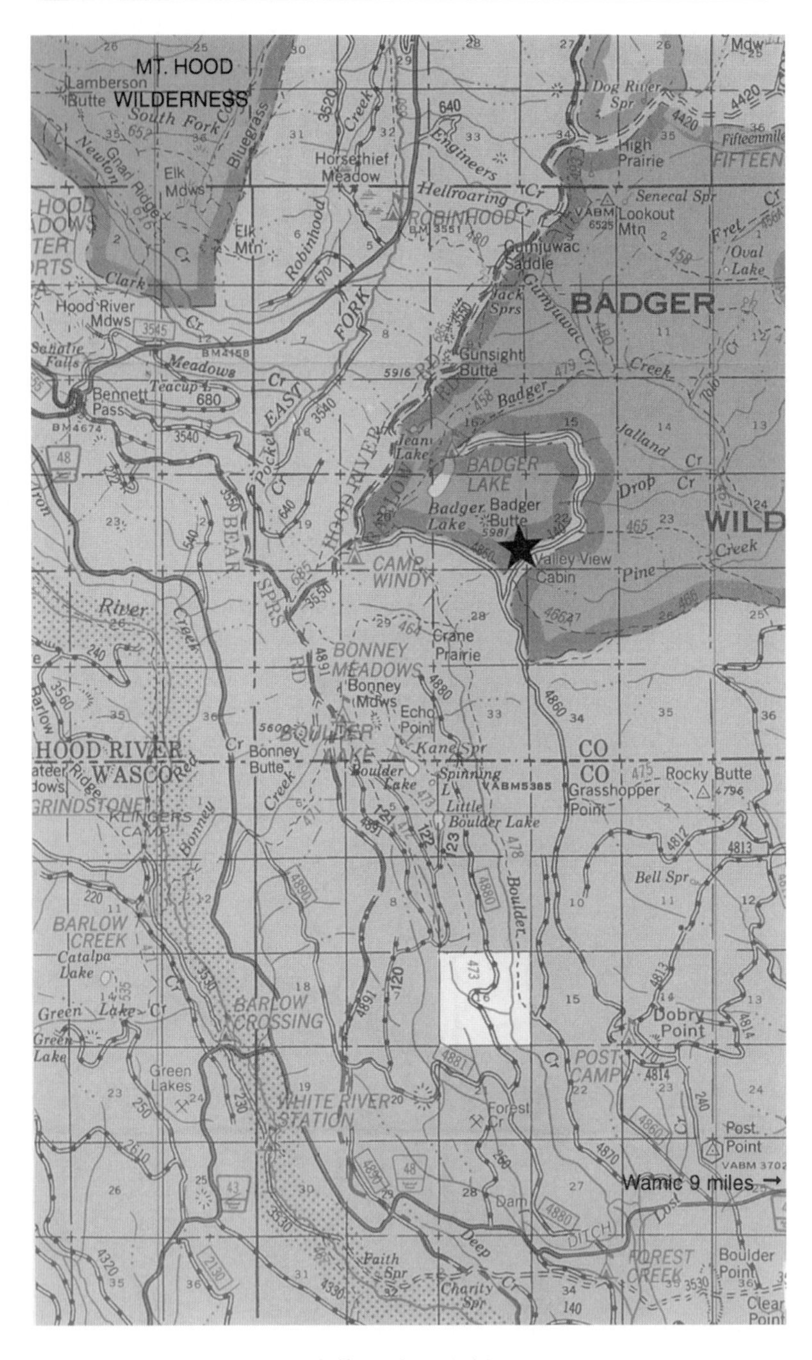

Valley View Cabin
Mt. Hood National Forest, Oregon

9 ⸙ Lost Lake Cabins

Mt. Hood National Forest

"I want to go soon and live away by the pond, where I shall hear only the wind whispering among the reeds. It will be success if I shall have left myself behind."

— Henry David Thoreau, *Journal*, December 24, 1841

Your Bearings

20 miles west of Hood River Ranger Dist. in Mt. Hood-Parkdale
30 miles southwest of the town of Hood River
50 miles southwest of The Dalles
90 miles southeast of Portland

Availability May through October, depending on snow conditions.

Capacity

A maximum of 15 people in the log cabin and in each of the two Adirondack shelters. However, a group half that size would be more comfortable. Ideal for families.

Description

One 15x15-foot log cabin, two three-sided Adirondack structures, and one hexagonal, canvas-covered structure. All are very pleasantly situated.

The cabin, which is wheelchair-accessible, consists of a single room with exposed log walls, shingle roof, and a lovely hardwood floor. The cabin also has massive hand-hewn beams overhead, two skylights, a vaulted ceiling, lakeside windows with folding wooden shutters, and a woodstove with a stone chimney. The firewood is stored in a cellar under the house but, ingeniously, can be retrieved by a trap door in the floor beside the stove.

Cost

Lost Lake Cabin and the two Adirondack shelters are each $50 per night. These structures have a *minimum* of two nights stay—three on holiday weekends. The entire area—the two Adirondack shelters, the log cabin and the canvas-covered hexagonal shelter—can be rented for $150 per night during weekends or $100 per night during the week. The structures accommodate a maximum of 45 people. There are parking spaces for 15 vehicles. A two-night deposit is required.

Reservations

These cabins are very popular, requiring reservations well in advance—for the busiest weekends in July and August, as much as 12 months ahead. For information and reservations, contact:

Lost Lake Resort and Campground
PO Box 90
Hood River, OR 97031
541-386-6366

How To Get There

From Portland, Oregon, travel east on Interstate 84 to the first Hood River exit (Exit 62). Travel through the town of Hood River on Oak Street to the stoplight at 13th Street. Turn right onto 13th Street and continue through town. After two miles the road name changes to Tucker Road. Stay on Tucker Road for two miles—to the Wind Master Corner intersection. Turn left—still Tucker Road—and continue for two miles to Tucker Bridge.

After crossing the bridge veer right (downhill) onto Dee Highway (State 281), and continue south along the West Fork Hood River to the town of Dee. Turn right onto Lost Lake Road, which is Forest Road 13. Continue on this paved but winding road for 15 miles to Forest Road 1340 and the Lost Lake Resort & Campground. Drive through the campground, following the signs for ORGANIZATIONAL CAMPGROUND. The cabins are about one mile beyond the entrance to Lost Lake Resort and Campground.

Elevation 3140 feet.

Map Location Mt. Hood National Forest, Township 1 South, Range 8 East, Section 15.

What Is Provided

There are four single beds—benches, really, affixed to the walls. They do not fold away, but they double as fine seats. Bring a sleeping pad or mattress of some kind—unless you don't mind sleeping on bare wood.

There is a fine hand-hewn picnic table inside also—though no cooking facilities apart from the barbecue and firering outside, where there is also a delightful porch with views of the lake.

In the area between the cabin and the Adirondack shelters is a vault toilet, a pleasant saunter away. About 100 yards from the cabin is a hexagonal structure with a canvas roof, and four picnic tables. This can be used by as many as 50 people as a gathering place.

What To Bring

Cooking utensils, and fishing gear. Since you can drive right to the rentals, take anything and everything you need to make your stay comfortable.

Setting

In almost every respect these three structures fail to fulfill our criteria for inclusion in this book, for they are neither remote nor secluded, and are adjacent to Lost Lake Resort, which has a store, boat launch (though, mercifully, no motorized boats are allowed) and, on the day we were there, seemed almost as busy as downtown Portland on a Friday afternoon.

Yet we have included these because they are unique. The Lost Lake Cabin was constructed in 1992 from existing Civilian Conservation Corps blueprints that date back to the 1930s. It was built as part of an ongoing series of workshops organized by Mt. Hood National Forest, and taught by experts from all over the country specializing in obscure but lovely trades, such as masonry and blacksmithing. Students and other participants pay tuition and fees of several hundred dollars to learn the myriad skills involved—everything from log- cabin construction to ornamental-hardware fabrication.

The two Adirondack shelters, three-sided with no front wall, were restored during a series of workshops, and the grand barbecue was built from solid stone by the masons and their students. It is most regrettable that the insides of these lovely structures have been almost completely covered with graffiti. Though structurally sound they are scarcely habitable in their present condition. One hopes after all the work that has gone into them that the Forest Service will be able to find a solution to this dismaying problem.

However, the Lost Lake Cabin, just beyond the Adirondacks, may be what the wise wilderness doctor ordered for the harassed Portland family. It is in immaculate condition, overlooks the lake, and is far enough away from the bustle of the campground and resort to allow a sense of seclusion and solitude. It is wheelchair-accessible. The cabin itself is remarkable in that it was built entirely without the use of electric tools. Even the ironwork was shaped by blacksmiths and their students on the site.

Lost Lake itself is stocked with tens of thousands of rainbow trout. There are also brook and brown trout and a few kokanee. There are several wheelchair-accessible areas for fishing at the lake, as well as a few wheelchair-accessible picnic sites, campsites and restrooms.

Around You

There are many old-growth trees in the vicinity. The Pacific Crest Trail is two miles to the southwest. It leads into Columbia Wilderness four miles northwest, and into Mt. Hood Wilderness ten miles southeast.

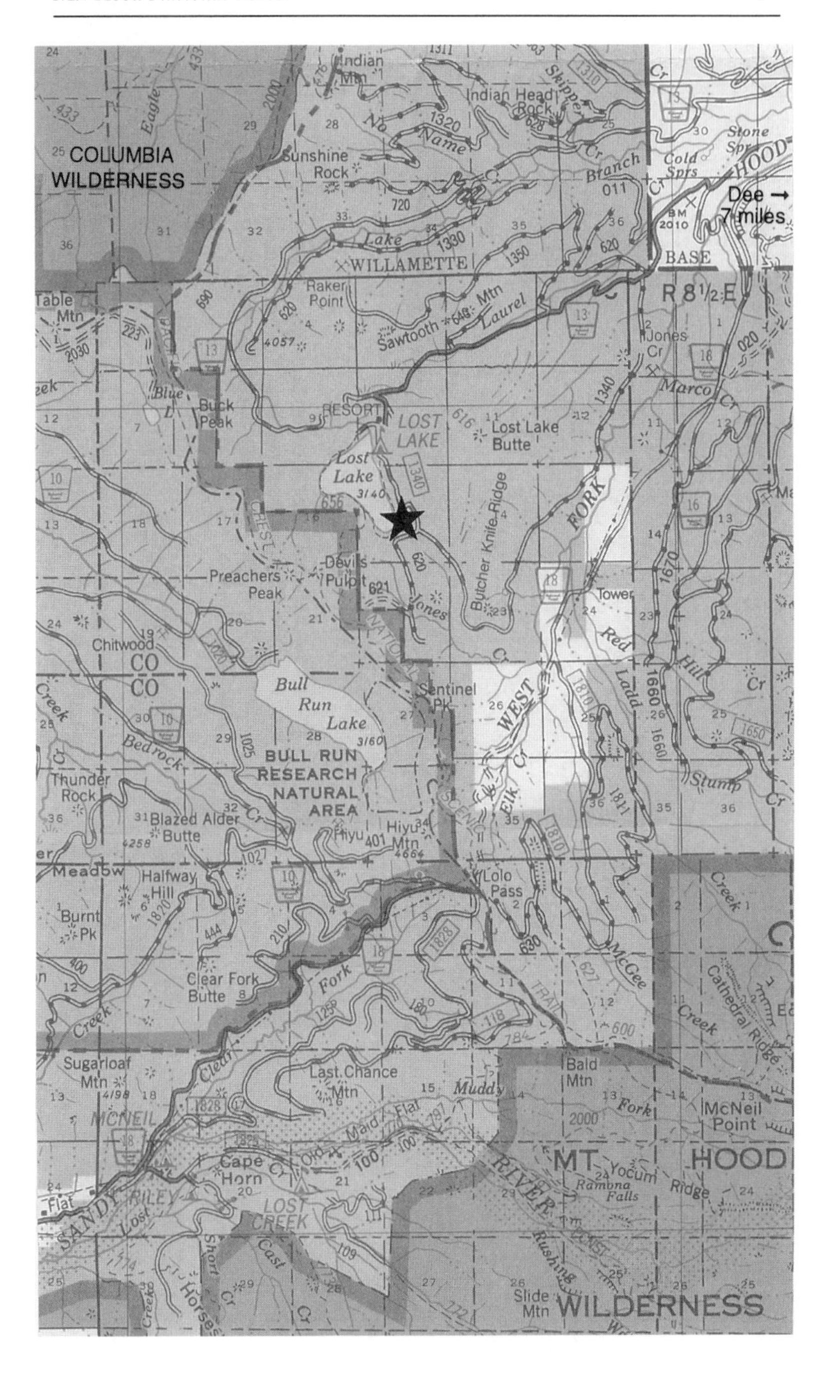

Lost Lake Cabins
Mt. Hood National Forest, Oregon

10 Warner Mountain Lookout

Willamette National Forest

"I never found the companion that was so companionable as solitude."

— HENRY DAVID THOREAU, *Walden*

Your Bearings

35 miles southeast of Oakridge
75 miles southeast of Eugene
140 miles southeast of Salem
140 miles southwest of Bend
190 miles southeast of Portland

Availability December 1 through May 10.

Capacity Four people maximum; two would be more comfortable. Not suitable for small children. No pets.

Description

14x14-foot room atop a 41-foot tower. 6x6-foot glassed-in observation cupola. 12x14-foot log cabin at ground level. Modern, with panoramic views. A perfect perch for winter weather watching.

Cost $25 for one night, $40 for two nights, and $20 per night thereafter.

Reservations

Available for as many as seven consecutive nights. For an application packet, maps, and further information contact:

Rigdon Ranger District
49098 Salmon Creek
Oakridge, OR 97463
541-782-2283

Safety Considerations

Always be well-prepared when entering the backcountry, especially on Warner Mountain in the wintertime. Carry tire chains and a shovel. The lookout is available as a *winter rental only*. During a typical winter you may be traveling across snow for 6-10 miles on skis or snowshoes to reach this remote and spectacular destination. The trip can be extremely difficult in snow and could take an entire day. Plan accordingly, and start early.

This ain't for the faint-hearted. Winter access offers an exciting challenge depending on snow conditions and your condition. The snowline can vary from 2800 feet to 4400 feet. This could mean a total vertical climb on skis or snowshoes of 1400 to 3000 feet. The final two miles are marked with orange poles to guide you across the upper meadows to the lookout. Consult the District Office regarding current road and snow conditions prior to your departure.

Because of the height of the lookout tower and the open catwalk, it might be risky to bring young children. Pets are not allowed. The stairway and catwalk are wooden and become slippery in rain, snow and ice. Occasionally, during strong winds, the tower will sway slightly. Don't worry, it's built to do this. The Forest Service advises that it is safer to remain in the tower than to attempt to descend the stairway during lightning or a wind storm. The lookout is well grounded.

How To Get There

From Oakridge, Oregon, travel three miles east on Highway 58. Turn right at Kitson Spring Road, toward Hills Creek Reservoir. After a few hundred yards, veer right onto Forest Road 21 (also called Rigdon Road). Remain on Road 21 as it crosses Hills Creek Reservoir at its southern end. From this point you are traveling a section of the historic Oregon Central Military Wagon Road. There is a 25-mile hiking trail along the banks of the Middle Fork of the Willamette River, which Forest Road 21 parallels. It passes through old-growth stands of Douglas fir, and alongside several campgrounds.

Continue south on Road 21 to its intersection with Forest Road 2129, or until you can drive no further due to snow. There is a small road sign on the right just before Road 2129 but it's easy to miss—at least it was for us. The sign at Road 2129 reads: LOGGERS BUTTE 14, MOON POINT TRAIL 10 but makes nary a mention of the lookout, though all along the way you will have noticed little lookout signs along the road every few miles.

Turn left onto Road 2129, also called Youngs Creek Road, and continue on its gravel surface until immediately after the 8 mile marker, where you reach Forest Road 439. Turn right here. After four miles you see another of those little lookout signs. Turn left; the lookout is less than 1/2 mile away.

Elevation 5800 feet.

Map Location Willamette National Forest, Township 23 South, Range 4 East, Section 29.

What Is Provided

Propane heat, propane oven, and propane light. Fuel is provided. Also a table and chairs, a single bed, fire extinguisher, shovel, and maps of the area. There is a 2-way radio available for *emergency communications only*. An outhouse is several hundred feet down the access road from the lookout.

This is a thoroughly modernized lookout, and should satisfy the whims of even the most thoroughly urbanized. It has fitted carpet, built-in fridge, stove with oven, built-in closets and shelving, built-in double bed, fly-screened windows, even a wood-framed screen door. The windows incidentally, are the modern, metal-framed, sliding type—a departure surely from the original design, though an understandable one.

On the ground there is a lovely log cabin with a shingle roof and, we're sorry to say, a concrete floor. It is available for storage. Even the vault toilet is of log construction, and has a shingle roof.

What To Bring

Drinking water is a must; or the means to treat snowmelt. Snow can be melted for your washing needs, but the Forest Service says that safe drinking water from snow cannot be assured. Extra food is a must—severe weather conditions may delay your departure. Prepare for harsh weather and pack accordingly.

History

This fire lookout was moved here to Warner Mountain from Grass Mountain. The cabin is a newly built replica of the old cupola-style lookout. The prototype for this style was constructed atop Mt. Hood in 1915. The cabin and cupola have 360° views.

Warner Mountain Lookout continues to serve as a summer home and duty station for the seasonal fire guards who play a key role in the detection of fires on the Rigdon Ranger District and adjoining forests.

Around You

To the south: Mt. Thielsen, Mt. Scott, and Mt. Bailey. To the northeast: the South and Middle Sisters of the Three Sisters, Broken Top and Mt. Bachelor. To the east: Diamond Peak and Diamond Peak Wilderness. To the northwest: Hills Creek Reservoir.

Moon Point Trail (3688) is well signed and can be reached from Forest Road 2129 or 439.

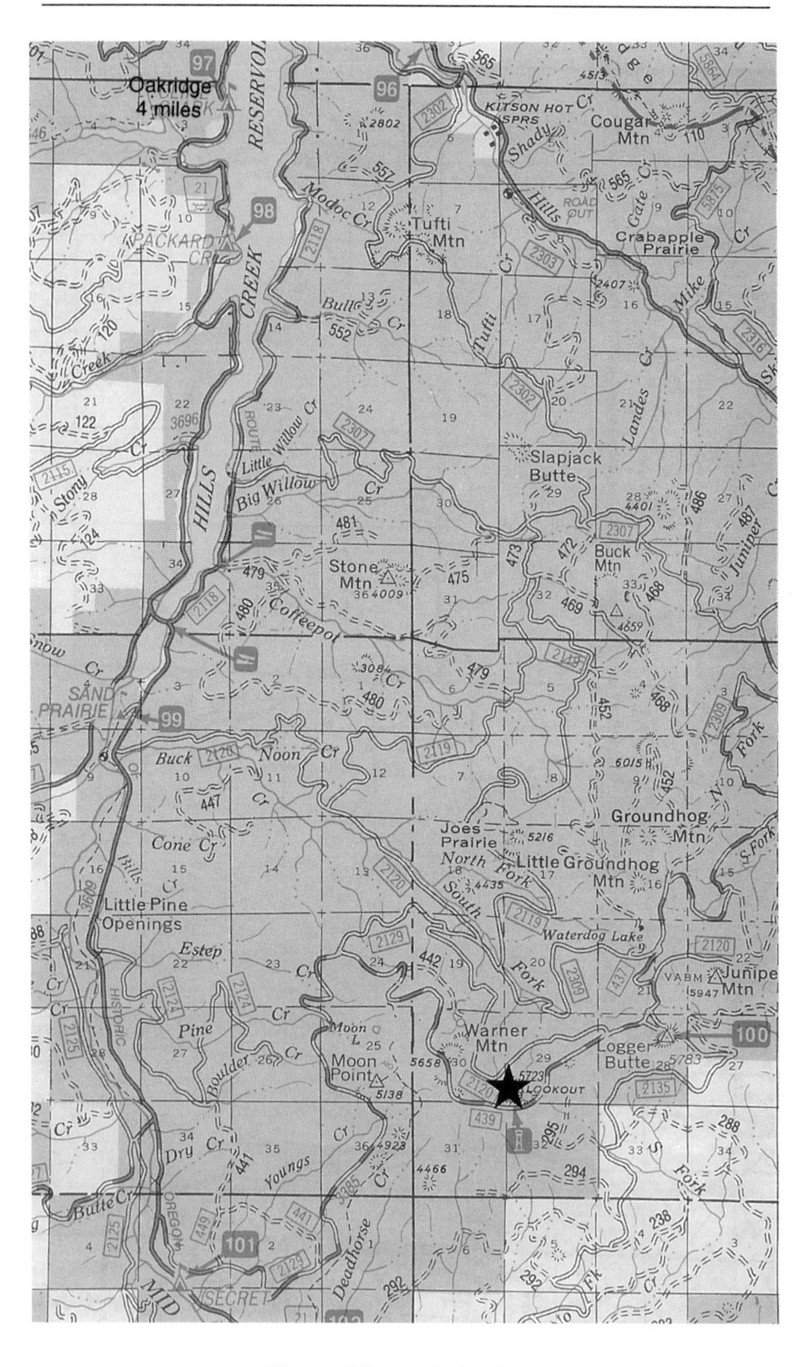

Warner Mountain Lookout

Willamette National Forest, Oregon

11 Indian Ridge Lookout
Willamette National Forest

"In the sun that is young once only,
Time let me play and be golden in the mercy of his means."

— Dylan Thomas, *Fern Hill*

Your Bearings

60 miles northeast of Oakridge
70 miles east of Eugene via Highway 126
95 miles east of Eugene via Highway 58 and Aufderheide Drive
100 miles west of Bend
135 miles southeast of Salem
185 miles southeast of Portland

Availability Late May to mid-November; actual dates depend on snow conditions.

Capacity One or two people—maximum of four. Not suitable for small children. No pets.

Description 16x16-foot room, all wood and windows, atop a 30-foot tower.

Cost $20 per night.

Reservations

Available for as many as seven consecutive nights. Check-out time is 1:00 PM. For an application packet, maps, and further information contact:

Blue River Ranger District
PO Box 199
Blue River, OR 97413
541-822-3317

Safety Considerations

Because of the height of the lookout tower, the open catwalk, and the location, it might be risky to bring young children. From the rocks around the lookout there are concealed vertical drops of a hundred feet or more and there are no guard rails. The stairway and catwalk are wooden and become slippery in rain, snow and ice.

Occasionally, during strong winds, the tower will sway slightly. Don't worry, it's built to do this. The Forest Service advises that it is safer to remain in the tower than to attempt to descend the stairway during lightning or a wind storm. The lookout is well grounded.

How To Get There

There are two routes from Eugene, one somewhat longer than the other, though we suggest the longer one because it will take you from south to north along the Robert Aufderheide Memorial Drive—and it is a memorable drive.

To take the Aufderheide route: from Eugene, travel 36 miles southeast on Highway 58 to the Westfir Junction. Turn left here and after two miles you will be at the start of Forest Road 19, the Aufderheide Drive.

Just before you reach the Aufderheide Drive—the Aufderheide National Scenic Byway, to unfurl its complete title—you may notice a red covered bridge over the river. This is known as the Office Bridge. It is, in fact, the longest covered bridge in Oregon, and it is unique in that it has a covered footwalk beside, but separate from, the roadway of the bridge itself.

Back when Willamette National Forest had the reputation of cutting more timber than any other national forest in the country, this section of road was the roadbed for the railroad line that was used to haul millions of trees out of the forest to the mill at Westfir. The tracks were replaced by this road after World War II.

Still on Road 19, you are now traveling parallel to the North Fork of the Middle Fork Willamette River. This is a long and lovely drive, green and watery, shadowed in places by old-growth trees, particularly at Constitution Grove, 27 miles north of Westfir Junction. A short trail takes you through the grove.

Continuing north and east through the Box Canyon area you will travel, first, parallel to Roaring River and, after that, the South Fork of the McKenzie River.

It is about 50 languid miles north from Westfir to Forest Road 1980. Turn left (northwest) here. Follow Road 1980 for 7.3 steep and winding miles of gravel until you reach signed Forest Road 247. Turn right. Continue for another 2.7 miles: there, on the left, you will see a locked green gate and a ROAD CLOSED sign. The lookout is less than 1/4 mile beyond the gate, although you can't see it until you get much closer.

The shorter route, for the brisk, brusque, or just plain busy, is to travel 45 miles east of Eugene on Highway 126 to Forest Road 19, three miles east of Blue River, and turn right. This is the northern end of the Aufderheide Drive. Continue south for 16 miles to Forest Road 1980, which will be on your right, just after French Pete Campground.

Turn right (northwest) onto Road 1980 for 7.3 steep and winding miles of gravel, until you reach Forest Road 247. It is signed. Turn right and continue for another 2.7 miles: there on your left you will see a locked green gate and a ROAD CLOSED sign. The lookout is less than a quarter of a mile beyond the gate, though you can't see it until you get much closer.

Forest Road 247 is quite narrow with precipitous, unguarded cliffs. Drive only in daylight, and very soberly.

Elevation 5405 feet.

Map Location Willamette National Forest, Township 18 South, Range 5 East, Section 18.

What Is Provided

Two nice beds, a table, two chairs, a small chest of drawers and some shelves. Unfortunately, it has no stove, no sink, no fridge, no heating, no fireplace, no light, and no water, but lots of firewood (for campfires), and an outhouse.

The day we were there, in July, there was a family very happily stoking up their own Hibachi on the rocks beside the tower.

What To Bring

Water. While most lookouts do not have inside potable water, this is the only one we have come across in our travels that is rented so bereft of what many would consider the basic essentials. If you are renting in early spring or late fall, good sleeping bags are a must, as are warm clothing and, of course, some form of lighting, heating, and cooking equipment.

History

Built in 1958, it is still used intermittently for fire detection during the summer months.

Around You

Three Sisters and Mt. Bachelor are to the east. Diamond Peak is to the south. Mt. Jefferson, Three Fingered Jack and Sand Mountain are to the northeast Mt. Hood and Mt. Washington are to the north, and the foothills of the Cascades are seen to the west.

Beneath you, off a sheer drop of several hundred feet, is a perfectly circular pond surrounded by tall trees.

Marring the spectacular view, somewhat, are three eyesores in the shape of three antennae that look like they were dropped there by angry aliens who had been refused Green Cards.

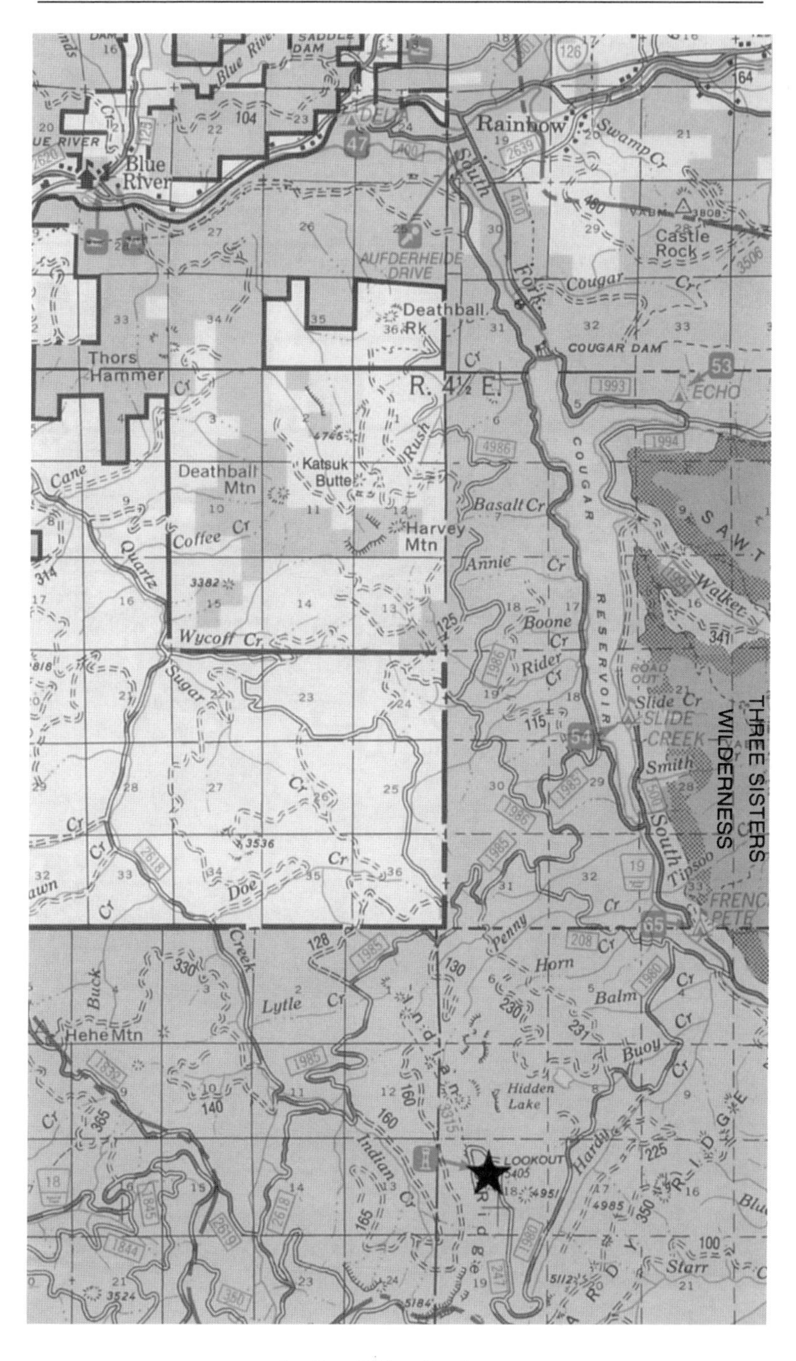

Indian Ridge Lookout
Willamette National Forest, Oregon

12 Box Canyon Guard Station

Willamette National Forest

"Our deepest fear is not that we are inadequate,
our deepest fear is that we are powerful beyond measure."

NELSON MANDELA, *1994 Inaugural Address*

Your Bearings

35 miles northeast of Oakridge
75 miles southeast of Eugene
105 miles west of Bend
135 southeast of Salem
185 miles southeast of Portland

Availability Mid-May through mid-November, depending on snow conditions.

Capacity Up to six people, but two adults, with perhaps one or two small children, would be much more comfortable.

Description 21x15-foot, two-room log cabin, in need of some T.L.C.

Cost $25 for the first night and $20 each additional night.

Reservations

Available for as many as seven consecutive nights. For an application packet, maps, and further information contact:

Blue River Ranger District
PO Box 199
Blue River, OR 97413
541-822-3317

How To Get There

Travel 45 miles east from Eugene on Highway 126, to Forest Road 19—which is five miles east of the Blue River Ranger District. Turn right onto Road 19. This becomes the Robert Aufderheide Memorial Drive. Continue south for about 26 miles. It is paved all the way. The guard station is across the road from Box Canyon Horse Camp.

Another route, and the one we recommend, is to travel 36 miles east of Eugene on Highway 58 to the Westfir Junction. Turn left here and in two miles you will be at the beginning of Forest Road 19, the Robert Aufderheide Memorial Drive. Box Canyon Guard Station is 33 miles from Highway 58 on Road 19.

The Aufderheide Drive, named for a former Supervisor of the Willamette National Forest, is one of the leafiest, greenest, shadiest, quietest, loveliest, wateriest drives you are likely to find. It follows the Wild and Scenic North Fork of the Middle Fork Willamette River most of the way to Box Canyon. We took this route in late July and saw fewer than a half-dozen cars the entire trip.

Box Canyon Guard Station is on the right, as you drive north, just past the sign for Skookum Creek Campground.

Elevation 3620 feet.

Map Location Willamette National Forest, Township 19 South, Range 6 East, Section 24.

What Is Provided

The kitchen/living room has a sink, faucets (not drinking water), a fold-out couch, woodstove, a two-burner-counter-top propane cook stove, closets, cupboards, chairs, and a table that has seen better days. There are no lights and no fridge.

The bedroom has two wooden single beds, and a separate entrance. There is an attached woodshed stocked with firewood for your use at the rear of the building. Some of the windows are fly-screened: all have shutters and curtains.

The cabin sits in a nicely fenced yard. There is a vault toilet, a picnic table, a firering and a round horse-corral, about 45 feet in diameter. You are welcome to use it, even for the kids. There is a Horse Camp across the road as well, built by the Civilian Conservation Corps in 1933. We were practically whinnying ourselves when we left.

What To Bring

All the water you need, or have the means to treat the local supply, since the water in the cabin is not potable, and in any event, is turned off at the onset of the first frost.

History

Box Canyon was named in 1880 by Charles McClane and Major Sears for its boxy shape. In 1934 the Forest Service built the guard station here to house a fire guard.

Around You

Beside the cabin, in the yard, is the trailhead for the McBee Trail (3523), which takes you into the Three Sisters Wilderness, 5.2 miles away. It also leads to the Crossing Way Trail (3307), 4.7 miles away, which in turn leads north to Roaring River Ridge or south to Three Sisters Wilderness. The Elk Creek Trail (3510), 9.5 miles away, can also be reached via the McBee Trail. Other wilderness trailheads can be found at Skookum Campground a few miles south at the end of Forest Road 1957. Across the road from the cabin in the Box Canyon Horse Camp is the trailhead for the Grasshopper Trail (3306). This leads to Chucksney Mountain, an area that the Forest Service calls, in lovely bureaucratese, "an undeveloped roadless recreation area."

There is a sign nearby on the road that has an interesting message:

SPOTTED OWLS

The forest west of this point is managed for four pair of northern spotted owls who prefer old growth to timber. The Forest Service has therefore protected some old growth habitat for the spotted owls. Within these areas, man's activities are restricted to reduce the impact on the foraging range and nesting areas of these birds.

A few hundred yards north on Forest Road 19, on the same side of the road as Box Canyon Guard Station, is Landess Cabin, set in a lovely meadow. This is a replica, built in 1972 by the Forest Service and a neighborhood youth corps, of the original cabin, which was built in 1918 by G. J. Landess and Smith Taylor for a Forest Service Fire Guard. It is not habitable.

About two miles south on Forest Road 19 is the trailhead for Trail 3567, which enters Waldo Lake Wilderness—less than 1/2 mile from the trailhead.

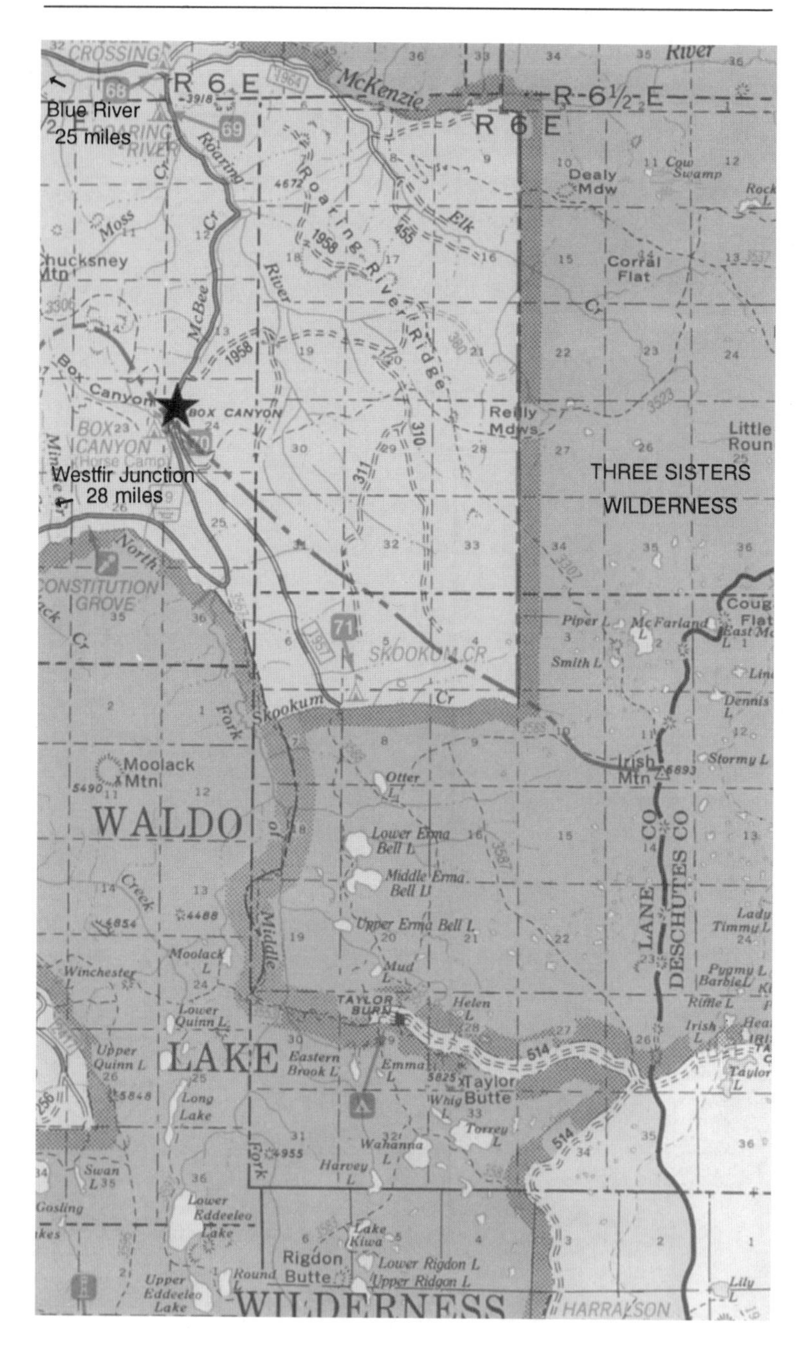

Box Canyon Guard Station
Willamette National Forest, Oregon

13 Acker Rock Lookout

Umpqua National Forest

"It isn't only the fact that you're alone, but the height and the amount of country you see makes it like being suspended between heaven and earth."

— Martha Hardy, Lookout Guard, 1945

Your Bearings

30 miles east of Tiller
60 miles east Canyonville
80 miles south of Roseburg
150 miles south of Eugene

Availability

June 23 through October 31.

Capacity

The lookout is designed to accommodate four people maximum, although for comfort we recommend no more than two. Not suitable for small children.

Description

12x12-foot cabin with narrow catwalk, perched high on a cliff. Spectacular views.

Cost $40 per night.

Reservations

Available for as many as three consecutive nights. For an application packet, maps, and further information contact:

Tiller Ranger District
27812 Tiller Trail Hwy.
Tiller, OR 97484
541-825-3201

When you visit Tiller Ranger Station to pick up your keys, we suggest you take the short trail to historic Red Mountain Lookout—relocated piece by piece to a small knoll on the grounds. Originally constructed in 1928 atop Red Mountain, this lookout is a fine example of the cupola-style design common in the 1930s. It was moved here in the autumn of 1985. It is not for rent.

How To Get There

From Tiller, Oregon, travel northeast on County Road 46 along the South Fork Umpqua River. This road parallels the river for 6.5 miles, where it becomes Forest Road 28, at the Forest boundary. You will be traveling a two-lane paved road through a lush forested canyon, leaving the broad Umpqua Valley behind. Along the way there are several campgrounds with access to the river, as well as waterfalls, hiking trails, and swimming holes.

Travel on Forest Road 28 for 12.2 miles. Turn right onto Forest Road 29 over a bridge across the South Umpqua River. At this point the road becomes a single-lane road with turnouts. Wonderful views of Acker Rock await you on the approach.

After 5.7 miles on Road 29, turn left onto Forest Road 2838, the turnoff for Acker Rock. This is a single-lane gravel road. After 1.6 miles turn left onto the 950 spur road, which is often gated and locked. Check with the Ranger Station to make sure you'll have clear access to the trailhead.

Unlock the gate and travel one mile on a steep, narrow road to the Acker Rock trailhead. The Acker Rock Trail (1585) is a half-mile, moderate-to-steep hike up to the lookout.

The trail cuts steeply through virgin Douglas fir forest. At the knife-edge ridgetop, climb the 27 wooden steps to the cabin's door. The fencing on the final stretch allows you to look ahead without worrying about your footing. Before the fence was built, all that existed on the precarious summit to guide your final access was a wire cable.

Elevation 4112 feet.

Map Location Umpqua National Forest, Township 29 South, Range 1 East, Section 13.

What Is Provided

Propane cook stove and fuel, lights and fridge, woodstove, single bed, table and chairs. An open-air pit toilet is nearby among the trees.

What To Bring

Drinking water is a must: it is not provided, though there is a spigot with potable water at the Tiller Ranger Station.

Setting

This lookout cabin anchored atop Acker Rock offers an expansive view of the upper South Umpqua watershed. The rock formation on which the lookout building sits is magnificent; sheer cliffs drop off for several hundred feet on the west and south sides.

Perched on an andesite cliff 2000 feet above the valley floor, you must be half mountain goat to relish this place. If you are queasy at heights, we recommend you choose a different rental. For those who think like a cloud, this is your spot.

The structure is creaky and weather beaten, but solid. A narrow catwalk around the lookout offers views of a patchwork of timber-producing landscapes including reforested clearcuts old and new. The foreground is dominated by rocks, trees and mountain air. The immediate area offers many large "thinking rocks" that are suspended over the valley—places to ponder gravity, to ponder balance.

The open-air pit toilet has views of sky and branches waving in the wind. We noticed a knotted pull rope leading up to it, via a steep but short trail. If you got this far, you probably won't need it.

History

Before the advent of airplane surveillance for fire detection, more than a dozen lookouts guarded the forests of the Tiller Ranger District. Acker Rock Lookout and Pickett Butte Lookout are the only ones that remain. Pickett Butte is available for rent also.

Around You

Panoramic view of landmarks along the Rogue-Umpqua Divide and magnificent peaks in the Rogue basin. On a clear day you can see the mountains in the Willamette, Rogue and Deschutes River watersheds. Several trails are maintained in the Tiller Ranger District for you to enjoy during your stay.

"So high you look down at the stars."

— Acker Rock Lookout Guest

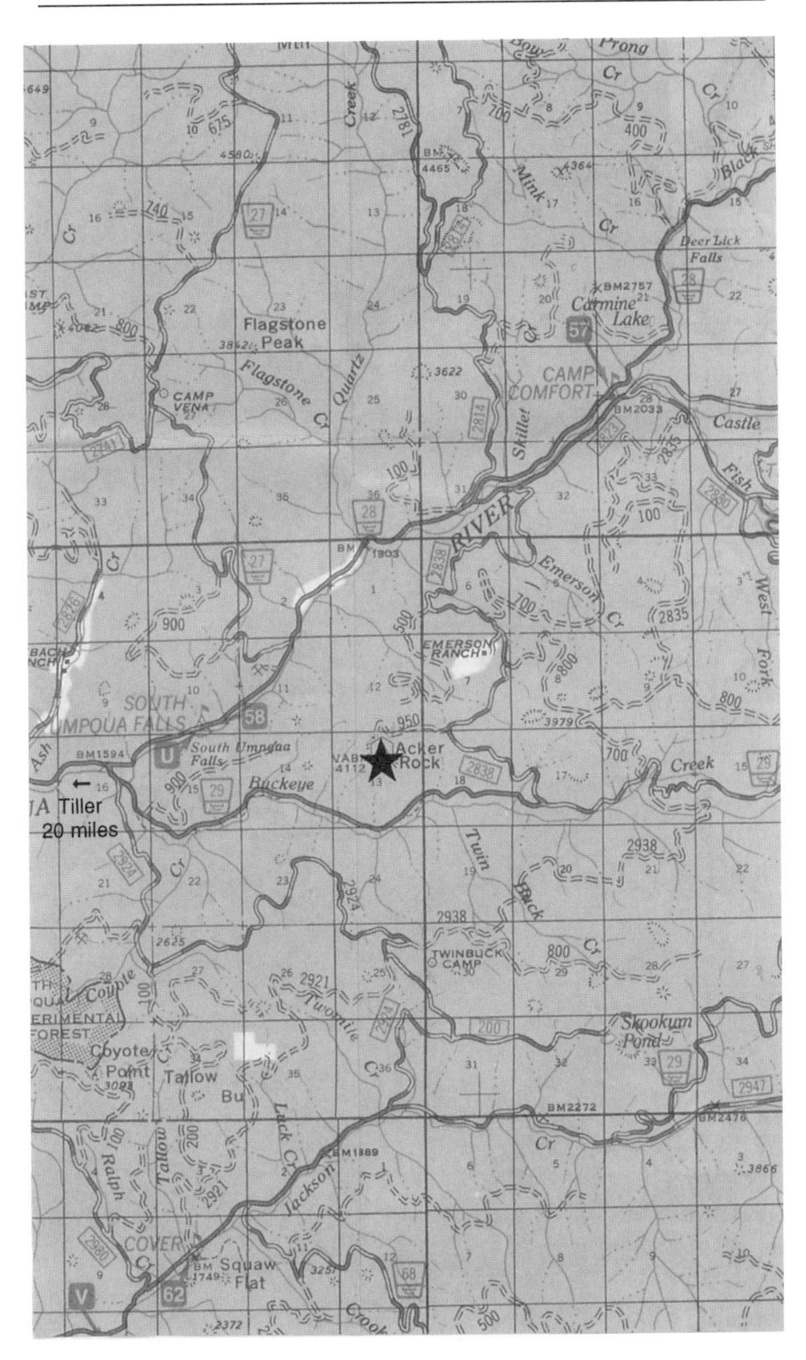

Acker Rock Lookout
Umpqua National Forest, Oregon

14 Pickett Butte Lookout

Umpqua National Forest

"A beautiful place—enjoyed our stay immensely. Great place for storytelling. Hope there wasn't a hidden tape recorder or that the walls don't talk."

— FROM A LOOKOUT'S GUESTBOOK

Your Bearings

10 miles northeast of Tiller
40 miles east of Canyonville
60 miles southeast of Roseburg
130 miles southeast of Eugene

Availability November through May.

Capacity

Maximum of four people—we suggest no more than two adults. The catwalk is narrow and the railing is low. Not suitable for small children.

Description 12x12-foot room atop a 41-foot tower.

Cost $40 per night.

Reservations

Available for as many as three consecutive nights. Reservations are from noon to noon. For an application packet, maps, and further information contact:

Tiller Ranger District
27812 Tiller Trail Hwy.
Tiller, OR 97484
541-825-3201

To initiate you to the lookout experience, visit the historic Red Mountain Lookout, relocated to the grounds of the Tiller Ranger Station in 1985. Now fully restored, it serves as an interpretive site you may enjoy when stopping by for maps or other information prior to your stay. This cupola-style lookout was originally built on Red Mountain in the upper Cow Creek Valley in 1928. It is not for rent.

How To Get There

From Tiller, Oregon, travel northeast on County Road 46 along the South Fork Umpqua River, and proceed three miles to Forest Road 3113, also called the Pickett Butte Road. Turn right, crossing the bridge. This turn is signed PICKETT BUTTE LOOKOUT 7.

Stay on Road 3113, the main gravel road, which climbs steadily upward. After 5.2 miles turn left onto Forest Road 300. At the intersection, signed PICKETT BUTTE LOOKOUT 2, veer left, staying on Road 300.

This narrow road takes a dip and then resumes its uphill grade. One mile up Road 300, you will come to a gate at the lookout's entrance. Past the gate, the road becomes steeper and narrower. It takes you directly to the site, 1/4 mile from the gate.

In winter, check with the Ranger Station prior to making the trip to find out whether the road is open. The Forest Service sometimes plows the access road for guests but be prepared for snowpack.

Elevation 3200 feet.

Map Location Umpqua National Forest, Township 30 South, Range 1 West, Section 29.

What Is Provided

The cabin is well-equipped with a propane cook stove and fuel, propane heater, lights, and fridge. It also has a table, chairs, stool, and single bed. The outhouse is at the end of the driveway.

What To Bring

Drinking water is a must: it is not provided. A spigot with potable water is available at the Tiller Ranger Station. Dress for extreme weather conditions and allow yourself extra food in case your departure is delayed by an unexpected storm.

Setting

It is thrilling to be 41 feet up on a wooden tower, some 57 steps from ground to front door. Negotiating the steep stairway is easier going up than down. The views are subtle, yet captivating.

History

Before the advent of airplane surveillance for fire detection, more than a dozen lookouts guarded the forests of the Tiller Ranger District. Today Pickett Butte and Acker Rock Lookout, which is also available for rent, are the only two that remain.

Around You

Overlooking the diverse landscape of the Tiller Ranger District, there are expansive views of the Rogue-Umpqua Divide, which is south and east of Pickett Butte, and has scenic peaks and landmarks capped with snow until early summer.

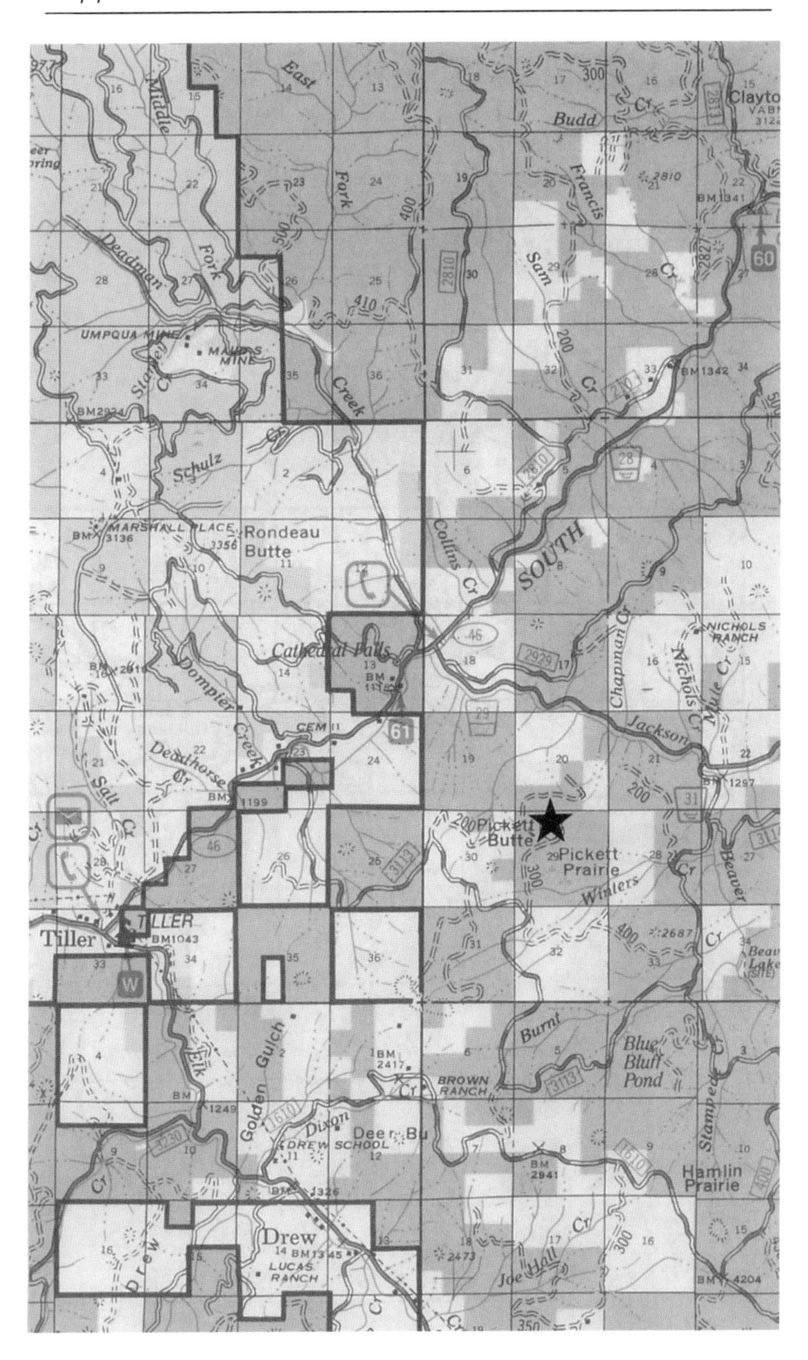

Pickett Butte Lookout
Umpqua National Forest, Oregon

15 Whisky Camp

Umpqua National Forest

"A place where human structures compliment nature's structures."

— WHISKY CAMP GUEST, SEPTEMBER, 1995

Your Bearings

20 miles east of Tiller
50 miles east of Canyonville
70 miles southeast of Roseburg
140 miles southeast of Eugene

Availability June through October.

Capacity Four people maximum. Ideal for families.

Description A lovely two-room cabin with covered front porch.

Cost $40 per night.

Reservations

Available for as many as seven consecutive nights. Check-in and check-out time is 12:00 noon. For an application packet, maps, and further information contact:

Tiller Ranger District
27812 Tiller Trail Highway
Tiller, OR 97484
541-825-3201

How To Get There

From Tiller, Oregon, travel northeast on County Road 46 along the South Fork Umpqua River for 6.5 miles to Forest Road 29. Turn right (southeast) over the bridge onto Road 29, which is a paved, two-lane road paralleling Jackson Creek.

After traveling 9.7 miles on Road 29, turn right and cross the Jackson Creek Bridge onto Forest Road 2925. The road becomes gravel at this point. (You might want to take a short side trip to the world's tallest sugar pine. The route is clearly signed).

Travel for 2-1/2 miles on Road 2925, then veer right onto Road 3114, signed WHISKY CAMP 4. Proceed 3.3 miles and turn left onto Forest Road 600. After 1/2 mile, at a junction, veer left again; it is signed WHISKY CAMP.

Continue another 1/2 mile to the gate, which is on the right, downslope of the road. You will have a key for the Master lock. The cabin is a short skip down the driveway.

Map Location Umpqua National Forest, Township 30 South, Range 1 East, Section 32.

What Is Provided

The two-room cabin has two sets of bunkbeds with mattresses, a propane heater and fuel, propane lights, and a fire extinguisher. Outside, there are picnic tables, a trash can, a metal firepit, and an unusually clean outhouse.

What To Bring

There is no water at the site so bring as much as you will need for drinking, cooking and washing. This is one of the few rentals without any pots or pans or eating utensils, so remember to bring your own. Bring clothing for extreme weather.

Setting

Quiet. It is a wonderful place to bring a family. The highlight of the cabin is its large, sheltered front porch. A perfect place for conversation or contemplation. The cabin, itself, nestles in a grove of incense cedar, facing southwest, and receives warm, dappled light throughout the day. A peeled-rail fence encircles the grounds. To the west of the cabin you will find a vault toilet and a shelter built by the Civilian Conservation Corps in the 1930s.

History

Experience what the life of a Forest Service fire guard was like before the days of roads and aerial fire detection. This cabin was the summer home of fire guards whose primary duties were to spot smoke, fight fires, give respite to the lookout guard, maintain phone lines and trails, and patrol the forest.

Around You

Just beyond the outhouse is a trail, built by volunteers, and open to equestrians and hiker. It leads to Beaver Lake, approximately five miles west. Or, you can drive to Beaver Lake from Road 31.

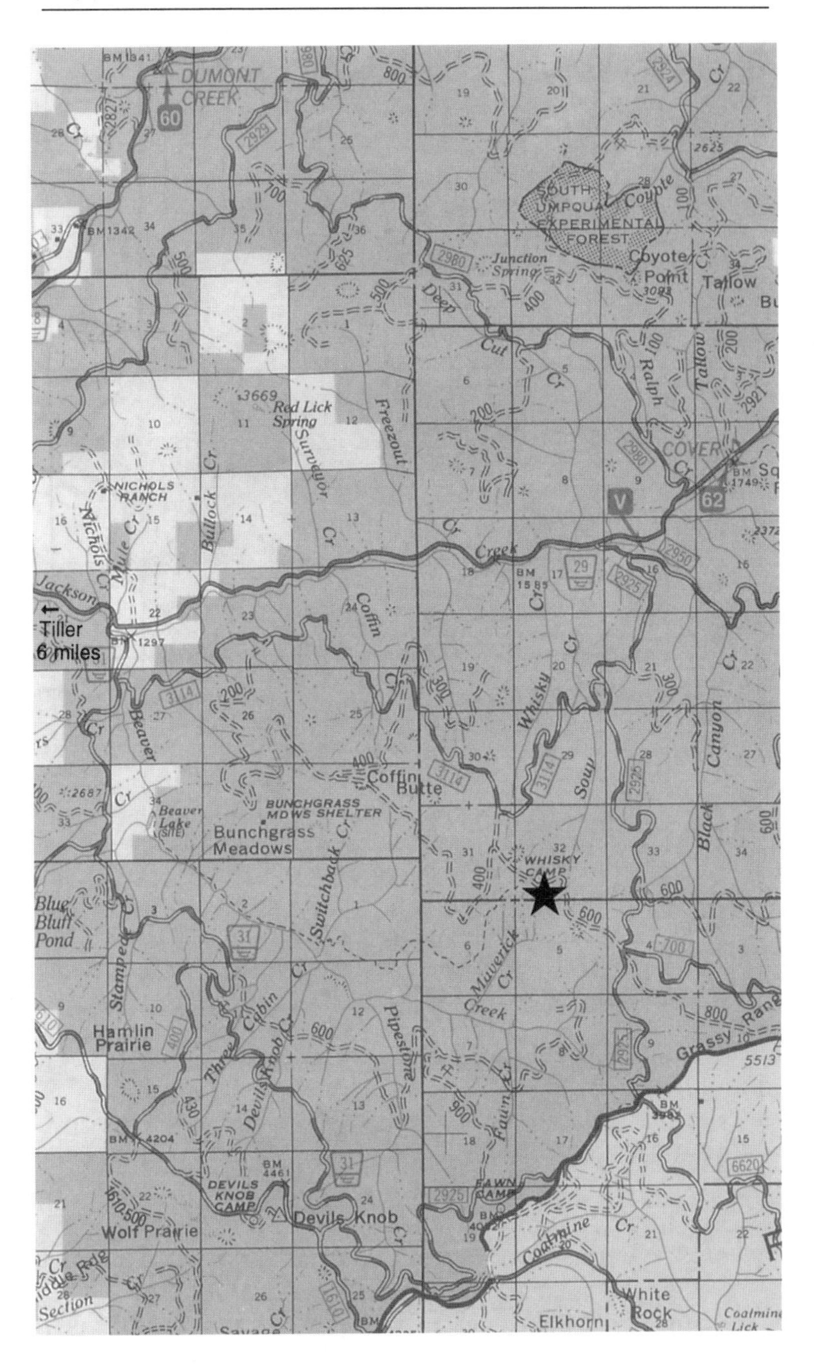

Whisky Camp
Umpqua National Forest, Oregon

16 Pearsoll Peak Lookout

Siskiyou National Forest

"I have never set foot on this mountain that I didn't get the most wonderful feelings I have ever felt. It is truly a unique place."

— From the Lookout's Guestbook

Your Bearings

25 miles northwest of Cave Junction
55 miles west of Grants Pass
85 miles west of Medford
100 miles west of Ashland
125 miles southwest of Roseburg

Availability June 1 through October 31.

Capacity A maximum of four people, though two would be more comfortable.

Description 14x14-foot room with large windows. Recently restored, with expansive views.

Cost Contact the Ranger District for current information.

Reservations

There is a Port-Orford-cedar road closure beginning October 1, requiring visitors to hike in seven miles via a trail from Onion Camp after this date. Available for as many as seven consecutive nights. For an application packet, maps, and further information contact:

Illinois Valley Ranger District
26568 Redwood Highway
Cave Junction, OR 97523
541-592-2166

How To Get There

There are two routes to the lookout—one by road, the other by trail. The route by road is open only from June to September, and then only to four-wheel-drive, high-clearance vehicles.

Beginning at the Illinois Valley Ranger District office in Cave Junction, Oregon, take Highway 199 north 8.6 miles to the Illinois River Road (County Road 5070) in Selma. Turn left (west) onto County Road 5070 and travel 10.9 miles from Selma to Forest Road 087, the McCaleb Ranch turnoff.

At the McCaleb Ranch turnoff, veer left and cross the low-water bridge over the Illinois River. Proceed straight across the private-property section of the road, paralleling the West Fork of Rancherie Creek on Forest Road 087. Travel 5.2 miles from the Illinois River crossing to Chetco Pass. This section is very slow and rugged.

At Chetco Pass, turn right and proceed 1.2 miles to Billingslea Junction, which is signed. At this junction, veer left. Travel 0.7 mile to a very small parking area on the right side of the road. From here the lookout is in view.

Travel time from the Illinois River to the parking area is approximately 1-1/2 hours. Do not attempt to drive the final 300 yards of road beyond the parking area. This section is very dangerous and there are no turn-around spots. Hike up the road until you come to the Kalmiopsis Wilderness Boundary. At this point the road becomes a trail. Summer temperatures often reach 100 degrees. The hike from the parking area takes only 35-45 minutes, but can be arduous in hot weather or high winds.

The alternative route is to backpack seven miles to the lookout from Onion Camp. To get to Onion Camp, travel 17 miles west on the Onion Camp Road (Forest Road 4201) from

Highway 199. The turnoff is four miles south of Selma and five miles north of Cave Junction. Park at the trailhead at Onion Camp and follow the Kalmiopsis Rim Trail (1124.2).

The trail passes over Eagle Mountain, where it becomes an old mining road. Proceed to Chetco Pass and from there hike the road to Billingslea Junction, 1.2 miles. Veer left at the junction and proceed to the lookout.

Elevation 5098 feet.

Map Location Siskiyou National Forest, Township 38 South, Range 10 West, Section 2.

What Is Provided

The table, chairs, bed, and cabinets have all been in this lookout since the 1930s. It also has a fire extinguisher, foot stool, oil lamp, and a map of the area. There is no stove or fridge nor any heat source.

In one of the drawers you will find a guestbook and a transcript of an interview with Albert Curnow, who staffed Pearsoll Peak Lookout in the 1930s. Tish did the interview more than a decade ago while working as an archaeologist for Siskiyou National Forest.

Also, note the two poems tacked to an inside wallpost. The first is *Mid-August at Sourdough Mountain Lookout,* by Gary Snyder (see page 34). The second is anonymous, and also a gem.

The outhouse is just below the cabin. It was airlifted to this remote mountaintop by helicopter, and placed to afford users the finest view.

What To Bring

No water is available so bring plenty. A spring is located about a mile down the main road from the lookout—watch for it on the way up. If you want to drink this water, use a filter or boil it. Bring a camp stove and fuel. Pack for hot days, and sharp, crisp summer nights.

Setting

Pearsoll Peak Lookout sits in a rugged and somewhat inaccessible section of the forest, just 197 feet outside the

Kalmiopsis Wilderness boundary. The lookout is perched on an exposed peak that has steep cliffs on all sides.

History

Mining activity in the Chetco Pass area during the Great Depression led to the construction of the first road into Pearsoll Peak. During World War II, the flat below the main saddle was used by the US Army as an enemy-plane detection site. Some evidence of this camp remains.

Pearsoll Peak is one of eight remaining lookouts in Siskiyou National Forest, and one of only two in the Illinois Valley Ranger District. Known as an L-4, this lookout was built in 1954, replacing an earlier cupola-style cabin constructed here in 1933. Prior to that Pearsoll Peak was the site of a primitive fire-lookout tent camp. The L-4 lookout kit could be ordered as a kit for $500 from Spokane, Washington, or Portland, Oregon. The L-4 models were also called "Aladdins" after their principal manufacturer. The lookout kit was designed with simplicity in mind; parts were numbered and blue prints were provided. The kit came ready to load on the backs of pack mules, the standard mode of transport into this and other remote Pacific Northwest mountaintops.

In 1973 the lookout was shut down, and was used thereafter only during periods of high fire danger. By the late 1980s, exposure and lack of use nearly led to its demise. But the story has a happy ending. Recognizing its charm and historic significance, the people of the Illinois Valley Ranger District, assisted by local volunteers and members of the Sand Mountain Society, united their skills and desires. Together, in 1991, they completely restored the lookout.

Now, in its full glory, the lookout is listed on the National Register of Historic Places and the National Register of Historic Lookouts. The site was dedicated in 1994 and has been available as a rental since then.

Around You

Kalmiopsis Wilderness Area, and exceptional views of Vulcan Peak, Chetco Peak, Eagle Mountain, Canyon Peak and, on a very clear day, Mount Shasta.

Port-Orford-Cedar Root Disease

The highly prized Port-Orford-cedar is native to southwestern Oregon and northwestern California. The tree is widely used for landscape plantings, hedges and windbreaks throughout the Northwest. A root fungus began killing ornamentals as early as 1922, and by 1952 had spread throughout most of the cedar's native range. Despite this, the watersheds of Rancherie Creek and its tributaries, en route to Pearsoll Peak Lookout, still have significant stands of healthy trees.

The fungus is spread by water-borne spores, making the risk of contamination greatest during the rainy season. Once infected, the trees die. You can help limit the spread of this disease by taking the following simple precautions required by the Illinois Valley Ranger District:

As a prerequisite to the issuance of your permit, before using the Pearsoll Peak Road, please wash your vehicle, especially the undercarriage and tires. There is a car wash at a service station in Cave Junction, or you may use the wash facility at the Illinois Valley Ranger Station, on the south edge of Cave Junction.

"All true paths lead through mountains."

— From the Lookout's Guestbook

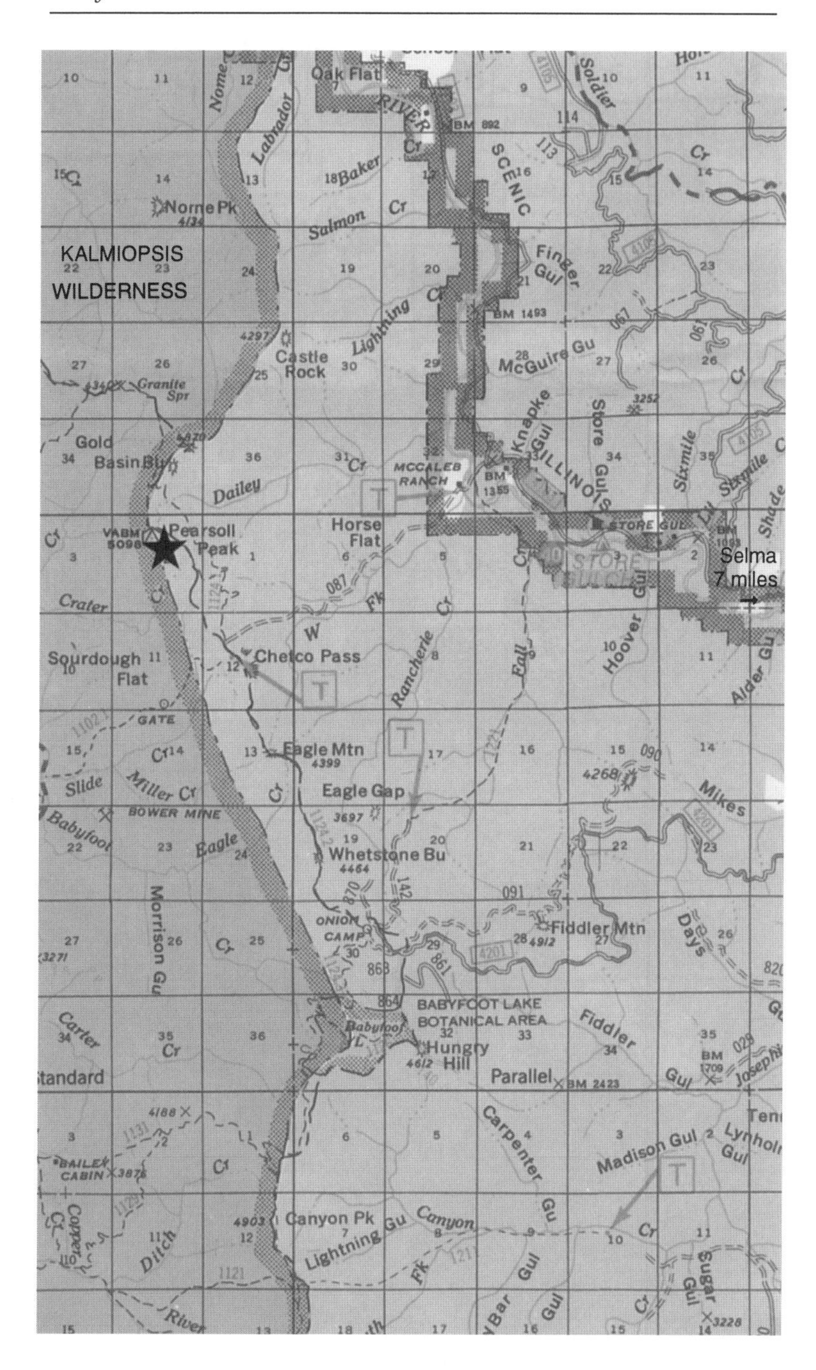

Pearsoll Peak Lookout
Siskiyou National Forest, Oregon

17 Snow Camp Lookout

Siskiyou National Forest

"I am he that walks with the tender and growing night, I call to the earth and the sea half-held by the night."

— WALT WHITMAN, *Song of Myself*

Your Bearings

25 miles southeast of Gold Beach
30 miles northeast of Brookings
60 miles north of Crescent City
180 miles west of Medford

Availability May to October.

Capacity A maximum of four people, though two would be more comfortable. Good for families.

Description 15x15-foot room with surrounding deck and railing, and 360° views.

Cost $30 per night.

Reservations

Available for as many as seven consecutive nights. For an application packet, maps, and further information contact:

Chetco Ranger District
555 Fifth Street
Brookings, OR 97415
541-469-2196

How To Get There

There are two well-signed, though distinctly different, routes to the lookout, both via the town of Brookings. One follows an inland route and the other follows the coastal highway, US 101. Each takes 1-1/2 to 2 hours driving time from the Chetco Ranger District office.

The advantage of the inland route is that you travel through the beautiful Chetco River Valley, though there are more miles of gravel roads than via Highway 101.

The advantage of the 101 route is that it travels 26 miles along the scenic coastal highway, offering spectacular views of the Pacific Ocean and its unique sea stacks just off-shore. There are several Oregon state parks along this drive with spectacular vista points.

Inland Route: From the Chetco Ranger District office, in downtown Brookings, Oregon, travel south on US Highway 101 for one mile to the North Bank Road (not shown on map). Watch for the sign NORTH BANK ROAD just before the bridge over the Chetco River. Turn left here. The North Bank Road makes an immediate right turn and heads upriver. The road is paved, winding and well-traveled.

After nine miles from Highway 101, you will reach the border of the Siskiyou National Forest. Here the road becomes Forest Road 1376. This is a single-lane, paved road, with turnouts and narrow bridges. It offers beautiful vistas of the Chetco River. The river is easily accessible via the picnic areas and campgrounds you pass along the way.

After 16 miles from Highway 101, Road 1376 crosses the South Fork Chetco River bridge. Continue on 1376 to Mile Post 18 where, at a fork, you will veer left toward Snow Camp—it is signed. After 3.1 miles from Mile Post 18, turn right, still on Road 1376. Proceed for another 5.5 miles to the lookout entrance—on your left at the yellow gate. Unlock the gate and go 1/2 mile to the lookout parking lot. A wheelbarrow is available to help haul your gear the 200 yards up a steep, rocky grade to the cabin.

US 101 Route: From the Chetco Ranger District office, travel 26 miles north on US 101 to the Hunter Creek Road. Turn right (east). Go five miles on the paved, two-lane county road, to where it becomes Forest Road 3680, and its surface becomes gravel. Go 18.5 miles on Road 3680 to its junction with Forest Road 1376. Turn right (uphill) onto Road 1376 for 1.2 miles to the yellow gate. Unlock the gate and proceed 1/2 mile to the lookout parking lot.

Along the route you may notice that the rocks in the roadcuts are shiny greenish-blue, and erode to reddish-orange soils. This is serpentine rock which was once on the ocean floor. Millions of years ago this rock was pushed upward, and it now caps many of the high mountain ridges in this region. It supports a unique community of rare trees and plants.

Elevation 4223 feet.

Map Location Siskiyou National Forest, Township 37 South, Range 12 West, Section 30.

What Is Provided

A double bed with a mattress, a table and chairs, counters and cupboards, a woodstove, and split wood stacked outside the door. In the center of the lookout is an Osborne Fire Finder. On the walls above the windows are beautiful, routed, wooden signs of key geographic place-names in the region. A picnic table and a chemical toilet are just outside.

What To Bring

No water is available so bring all you will need.

Setting

Overlooks the Pistol River Drainage, the 180,000-acre Kalmiopsis Wilderness, the Big Craggies Botanical Area, and, incidentally, the Pacific Ocean—which of course glistens in the moonlight. With light pouring in from all sides, you'll soon understand how an eagle feels in its nest.

Adjacent to the lookout is a communications building and tower. Though a bit of an eyesore, they do not significantly obstruct the view.

History

Snow Camp Lookout was built in 1958, and was used as a lookout until 1972. It was then reincarnated in 1990 as a recreational rental property. The first lookout on this site was built in 1924 and was used during World War II to detect enemy aircraft.

In the morning of September 9, 1942, Nobuo Fujita became the only man to make an enemy bombing run over the continental United States. Fujita placed his Will and a lock of his hair in a box to be sent back to Japan. He then put his four-hundred-year-old samurai sword aboard a single-engine float plane, and took off from submarine 1-25 off the coast of Cape Blanco, Oregon, convinced he was facing certain death. He flew 50 miles southeast over a wooded area where he dropped one incendiary bomb on the slopes of Wheeler Ridge. He saw flames spreading through the trees, flew on, then dropped another bomb before flying back to his submarine.

His attack was intended to start a massive forest fire that would strike panic into the heart of America, and serve as a counter-strike to the U.S. bombing raid on Tokyo the previous April. However, a fire-lookout guard spotted the smoke and alerted a fire crew, and the fire was soon brought under control.

Three weeks later, Fujita launched another attack, this time on Grassy Knob outside the coastal community of Port Orford, but due to wet conditions that year, it too fizzled. Twenty years later he returned to Brookings as an honored guest, bringing with him the same 400-year-old samurai sword, a traditional Japanese symbol of peace and friendship. The sword is on display at the Chetco Community Library, 420 Alder Street, Brookings, Oregon.

Around You

Snow Camp Trail (1103) is accessible from the parking lot. It offers an intimate look at the unique geology and plant communities of this area.

Fairview Meadow, just northwest of the lookout on Trail 1103, is thought to be the result of burning by Native Americans and early settlers despite the fact that this area gets up to 100 inches of rainfall annually. But it is lovely result.

In Fairview Meadow is the Enchanted Forest, a hauntingly beautiful stand of old-growth Douglas fir. The lichens hanging from the trees take all their nutrients and water from the air.

To hike to Snow Camp Meadow, take Trail 1103 from the parking area. If you keep your eyes peeled, you may see elk, deer, bear, and blue grouse. Look for western tree frogs and caddis flies in the pond at Snow Camp Meadow.

About 1/2 mile east of the lookout you will find a bog which is home to the cobra plant, *Darlingtonia californica.* It survives by trapping insects inside its tubular leaves and digesting them with the aid of bacteria, to provide itself with nutrients such as nitrogen, which is not available in the serpentine soils.

The lookout itself has been through several reincarnations and may have more to go. Take good care of it.

"Nobody sees a flower — really — it is so small it takes time — we haven't time — and to see takes time, like to have a friend takes time."

— Georgia O'Keeffe

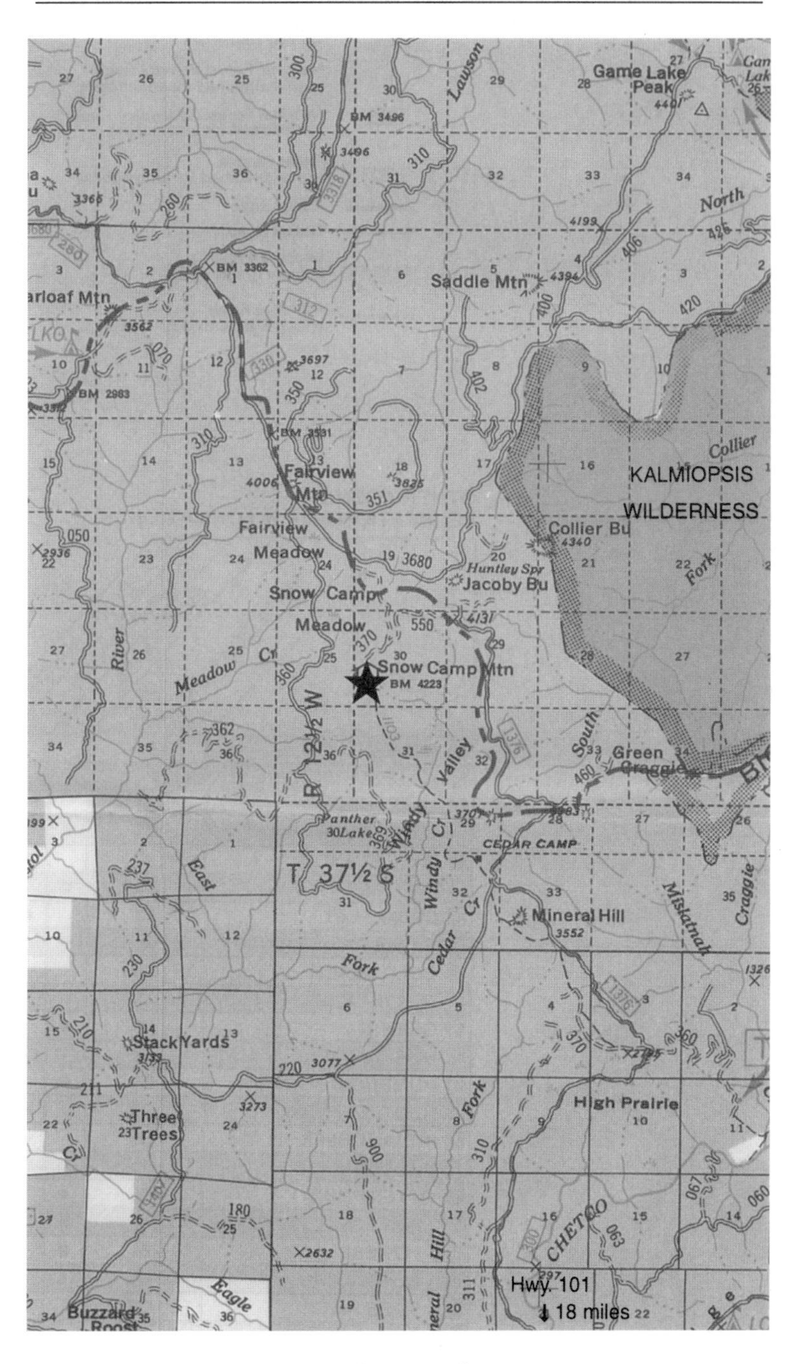

Snow Camp Lookout
Siskiyou National Forest, Oregon

18 Packers Cabin

Siskiyou National Forest

"...beautiful setting and the warmth of fellow travelers gone before."

— FROM THE CABIN'S GUESTBOOK, DECEMBER 1994

Your Bearings

25 miles northeast of Brookings
35 miles southeast of Gold Beach
50 miles north of Crescent City
225 miles southwest of Eugene

Availability Year-round, weather permitting.

Capacity

Up to a dozen people, though that would be taking the cabin's name a bit too literally—four to six would be much more comfortable. Ideal for families.

Description

Historic and charming, 28x16-foot, 3-room, cabin with wooden porch and railing. Wheelchair-accessible.

Cost $20 per night.

Reservations

Available for as many as seven consecutive nights. For an application packet, maps, and further information contact:

Chetco Ranger District
555 Fifth Street
Brookings, OR 97415
541-469-2196

How To Get There

From the Chetco Ranger District office, in downtown Brookings, Oregon, travel south on US Highway 101 for one mile to the North Bank Road (not shown on map). Watch for the sign NORTH BANK ROAD just before the bridge over the Chetco River. Turn left here. The North Bank Road makes an immediate right turn and heads upriver. The road is paved, winding and well-traveled.

After nine miles from Highway 101, you will reach the border of the Siskiyou National Forest. Here the road becomes Forest Road 1376. This is a single-lane, paved road, with turnouts and narrow bridges. It offers beautiful vistas of the Chetco River. The river is easily accessible via the picnic areas and campgrounds you pass along the way.

After 16 miles from Highway 101, Forest Road 1376 crosses the South Fork Chetco River Bridge. Continue one mile west to Forest Road 1917. Turn right onto Road 1917 and travel six more miles northeast on this narrow, moderately steep, gravel road. A sign indicates this road is not accessible to trailers, though it is suitable for passenger cars.

Stay right on Forest Road 1917 at its junction with Road 060. The driveway to the cabin is 2.5 miles ahead on the right side of the road. The only other destination on Forest Road 1917 is the Quail Prairie Lookout, 3.5 miles (according to the sign) east of Packers Cabin.

Elevation 2073 feet.

Map Location Siskiyou National Forest, Township 39 South, Range 12 West, Section 23.

What Is Provided

Woodstove and firewood, folding chairs, benches, shelves, and a large wooden table. The guestbook is a treasure. There are two small bedrooms. One is furnished with three sets of bunkbeds stacked three high, sleeping a total of nine. The other bedroom has a double bunk and a single bunk.

Outside you will find a vault toilet, picnic tables and a firering. The cabin and outhouse have ramps, wide doorways, and special door handles which make it fully accessible for wheelchair users.

What To Bring

Drinking water. There is a natural spring at the site, but the Forest Service recommends you use a filter or purifying system if you want to sample its waters.

Setting

This historic outpost is sequestered on the edge of a forest; perfect for artists needing a place to just be. On one side of the yard is a thick cluster of tall Douglas-fir trees, on the other, mature tanoak. To the north there is a natural spring set in a small meadow surrounded by low brush.

History

The cabin was built about 1930 as field headquarters for the backcountry packers who, on horseback or by mule train, regularly supplied the lookout guards and field crews with food and equipment. The structure was renovated in 1990 with funds from a Regional Forester Challenge Grant and the help of volunteers from Brookings and Harbor.

As you sit in the front yard on Long Ridge at Packers Cabin, listening to the water gurgle at the natural spring and the acorns fall, you are playing your part in maintaining the long history of this well-loved cabin in the Siskiyou Mountains.

Around You

Quail Prairie Lookout, at an elevation of 3033 feet, is, according to the wooden sign, 3-1/2 miles east of Packers Cabin, on Forest Road 1917. This lookout is staffed from July to October. You are welcome to visit the site and enjoy the view from its 50-foot tower: the Chetco River Valley, Big Craggies and part of Kalmiopsis Wilderness.

In the natural openings of the forest watch for Roosevelt elk, blacktail deer, black bear, wild turkey, quail, blue grouse, and red-tailed hawks. The best times are, of course, the early morning hours and at twilight.

For mountain bikers, the Long Ridge Loop (8.6 miles), along Forest Roads 060 and 1917, offers a scenic trip on graveled surfaces.

"The stars shine so brightly it's as if they'll never go out."

— From the Cabin's Guestbook

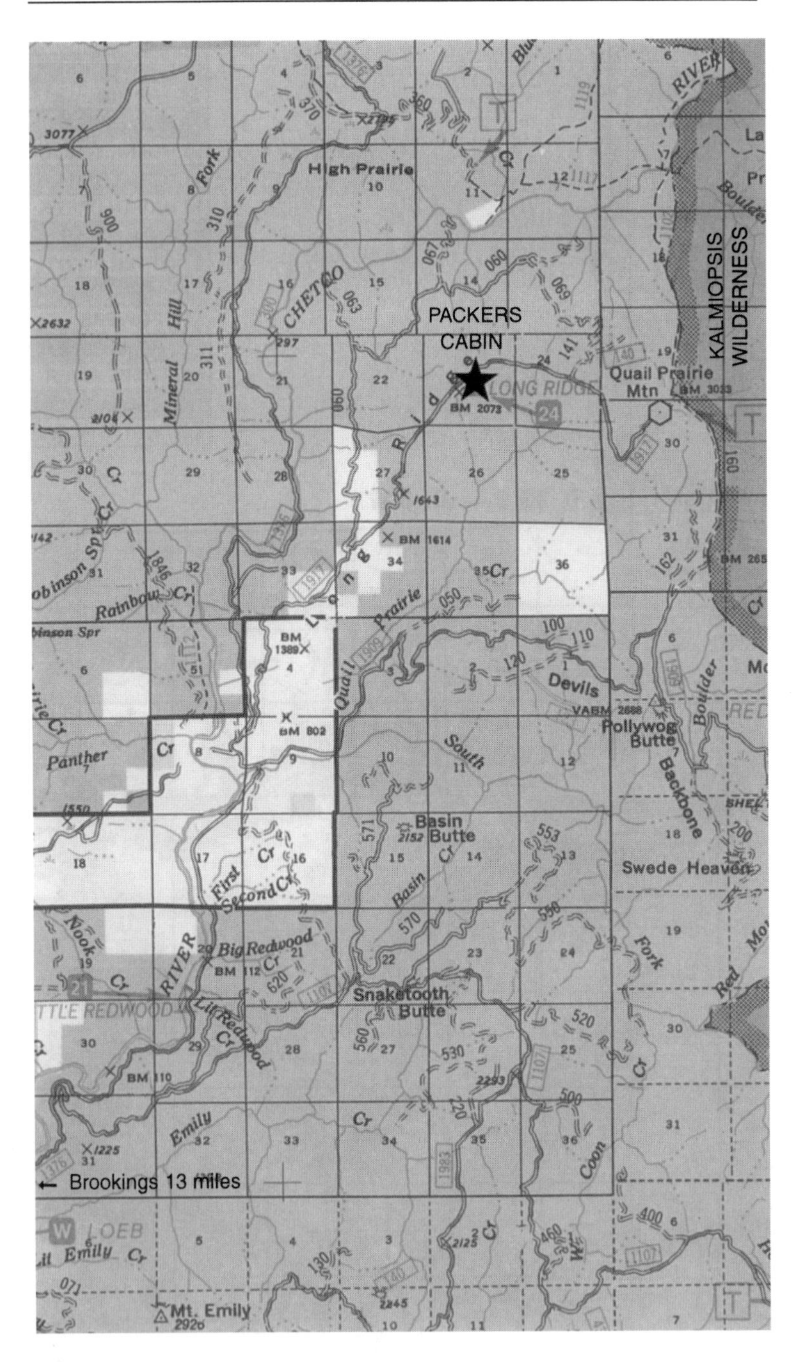

Packers Cabin
Siskiyou National Forest, Oregon

19 Ludlum House

Siskiyou National Forest

"We want to rest, we need to rest and allow the earth to rest. We need to reflect and to discover the mystery that lives in us . . ."

— U.N. Environmental Sabbath Program

Your Bearings

20 miles southeast of Brookings
50 miles southeast of Gold Beach
50 miles north of Crescent City
170 miles west of Medford

Availability Year-round, weather permitting.

Capacity The house can comfortably hold eight people, and less comfortably, as many as fifteen. Ideal for families.

Description Historic, remote, two-story, 3-room house with front porch, railing, and Port-Orford-cedar shake siding.

Cost $20 per night.

Reservations

Available for as many as seven consecutive nights, from noon to noon. For an application packet, maps, and further information contact:

Chetco Ranger District
555 Fifth Street
Brookings, OR 97415
541-469-2196

How To Get There

The route to Ludlum House is a paved, two-lane, well-traveled byway most of the way. Go south from Brookings on US 101 for eight miles to the Winchuck River Road (County Road 896). Turn left (east), and after six miles Road 896 becomes Forest Road 1107. Go an additional 2.3 miles to the junction of Road 1108. Veer left and uphill, onto Forest Road 1108. It too is paved at first, though it narrows and turns to gravel along Wheeler Creek. The driveway of Ludlum House, which is on the right side of the road, is 2-1/2 miles from the 1108 junction. The house is 200 yards down the driveway.

Map Location Siskiyou National Forest, Township 41 South, Range 12 West, Section 3.

What Is Provided

This homestead lies in a sunny, open meadow at the edge of the forest. Its large, sheltering front porch is a lovely spot to listen to the creek and the birds. Inside the house are benches, woodstove, work surfaces for food preparation, and a picnic table. The upstairs area is divided into two rooms. There are no beds, no electricity and no firewood or drinking water. Outside you will find an outhouse, several picnic tables and a firering.

What To Bring

Drinking water and firewood Bring along your own folding chairs for porch-sitting.

History

Built in the 1930s—by most accounts in 1937 or 1938—the cabin was originally a homestead. The last private owner of the house and property, and the person from whom it derives its name, was Robert C. Ludlum. He was an oil company executive who, in 1951, quit his job in Japan, returned to the United States, and, with his wife and two children, moved into the house.

The family moved to California the next year, though Ludlum returned to the Winchuck from time to time to work on his place and spend time here. He eventually sold it to the Forest Service in 1969.

Over the years, Ludlum House has developed a loyal group of visitors and guests who return to the cabin again and again. It has been the site of weddings, baptisms, reunions, and many other of life's significant events. In recent years it has twice served as a base for crews battling major forest fires in the area: the July Fire in 1986 and the Chrome Fire in 1990.

We understand from Siskiyou National Forest personnel that Ludlum House is scheduled for thorough renovation in 1996-1997.

Around You

Amid groves of myrtle and redwood, this homestead is in the Coast Range of the Siskiyou Mountains in southwestern Oregon. The Pacific Ocean is just a 30-minute drive away.

A mature Oregon myrtle grove shades the cabin's large parking area—which is suitable for passenger cars and recreational vehicles. Several picnic tables surround the campfire pit. Other picnic tables are scattered amid the trees. The front porch is just 100 feet from Wheeler Creek.

The confluence of Wheeler Creek and the Winchuck River is a short walk from the driveway gate, down the spur road. A sunny, gravel beach makes a fine spot to enjoy the water and the scent of the myrtle trees. The Oregon myrtle, from which we get the culinary bay leaf, often grows in clumps of five to ten stems, all coming from one root.

The Winchuck River is renowned for its winter-run steelhead, though fishing is allowed downstream of Wheeler Creek only. Licenses are available in Brookings sports shops or from the Oregon Department of Fish and Wildlife.

Wheeler Creek is a protected spawning ground for fall chinook and winter steelhead, and a year-round home to cutthroat and rainbow trout. Typically, chinooks start spawning here in October, depending on when the autumn rains begin.

Steelhead can be spawning here as late as January. If you do see fish spawning, it is important not to disturb them.

To ensure a healthy population, the Forest Service asks anglers to release any fish measuring less than eight inches. These small fish are smolts, and if released will come back to this creek as adults. In their efforts to improve fish habitat, the Forest Service has installed a variety of devices in the river to restore the natural mix of pools and riffles in the stream.

On the hill directly across the road from Ludlum House is a stand of old-growth redwoods. Though there is no trail, the steep hike is a rewarding one. Just across Wheeler Creek is an old apple orchard which still produces delicious apples in the fall. It is a good place for a picnic any time of year.

Fish carving in back of Ludlum House.

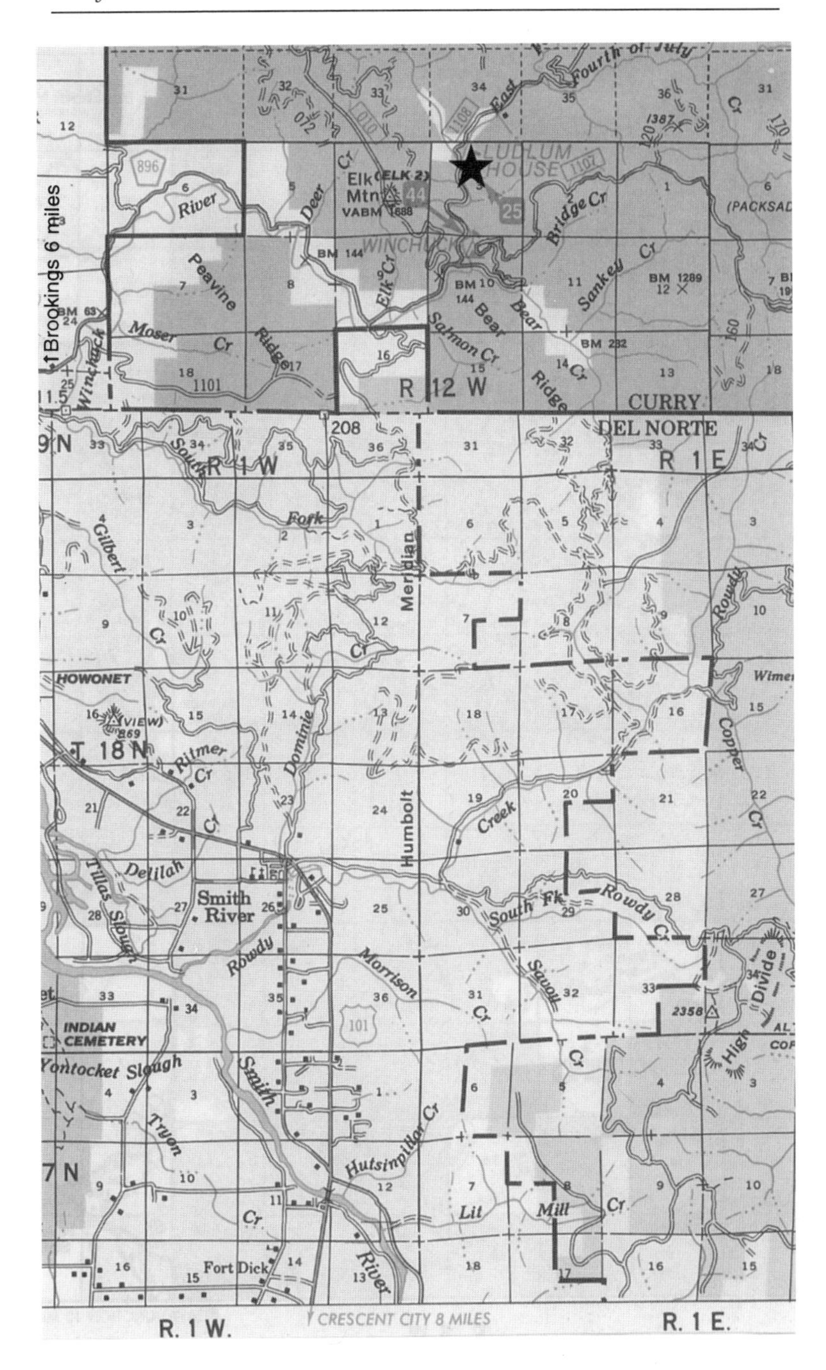

Ludlum House
Siskiyou National Forest, Oregon

20 Ditch Creek Guard Station

Umatilla National Forest

"I am no more lonely than a single mullen or a dandelion in a pasture, or a bean leaf, or sorree, or a horsefly, or a bumblebee."

Henry David Thoreau, *Walden*

Your Bearings

25 miles west of Ukiah
25 miles southeast of Heppner
75 miles southwest of Pendleton
75 miles west of La Grande
110 miles northwest of John Day

Availability Year-round.

Capacity Four people maximum.

Description One-story cabin with bedroom, living area, bathroom and kitchen.

Cost $35 per night, plus a $50 refundable deposit.

Reservations

Available for as many as seven consecutive nights. For an application packet, maps, and further information contact:

Heppner Ranger District
PO Box 7
Heppner, OR 97836
541-676-9187

How To Get There

From Heppner, Oregon, travel one mile south toward the Willow Creek Reservoir on Willow Creek Road. Turn left (east) onto the Blue Mountain Scenic Byway (Highway 678—which becomes Forest Road 53 at the National Forest boundary). Continue on Road 53 to its junction with Forest Road 21, a distance of four miles. (It is 21 miles from the town of Heppner to Forest Road 21). Turn right on Road 21 (gravel), and travel three miles south. The Guard Station is on a small hill on the west side of the road, up a short drive.

From Pendleton take Highway 395 south for 47 miles to Highway 53 West—about a mile or so west of Ukiah, Oregon. From Ukiah take County Road 244 west (it becomes Forest Road 53) for 22 miles to Forest Road 21. Turn left (south) on Road 21 and travel three miles on gravel to the Guard Station.

Safety Considerations

The portions of Willow Creek Road and Forest Road 21 closest to the Guard Station are not plowed in the winter, so the nearest parking from October through May, depending on snow accumulation, could be up to 5-1/2 miles away, at Cutsforth Park. Contact the Heppner Ranger District for current road and weather information prior to your visit.

Elevation 4800 feet.

Map Location Umatilla National Forest, Township 5 South, Range 28 East, Section 21.

What Is Provided

Four single beds, propane heater and fuel, water-heater, propane cook stove and fridge, eating utensils, pots and pans, fire extinguisher, cleaning supplies, smoke and propane alarms, and maps of the area. The firepit outdoors may be used only during periods of low fire risk. During the winter, the water is turned off to prevent pipe-freeze.

What To Bring

Bring enough water for your drinking and cooking needs, or have the means to treat the local water. Bring firewood for use in the outdoor firepit.

History

The building complex was built in 1934, and is a National Historic Landmark, exemplary of the architectural style used by the Civilian Conservation Corp during the 1930s in the Pacific Northwest.

Around You

Outside the guard station is a small pond where anglers can expect to catch brook and rainbow trout. The Heppner Ranger District has done an excellent job of compiling maps and information—all included in your rental packet. The area offers 26 miles of hiking, skiing, and horse trails.

Penland Lake, at an elevation of 4950 feet, is less than four miles east on Road 2103, and offers trout fishing, boating and picnicking. There are five tent sites, five picnic sites, and a boat launch, but no potable water or garbage pick-up.

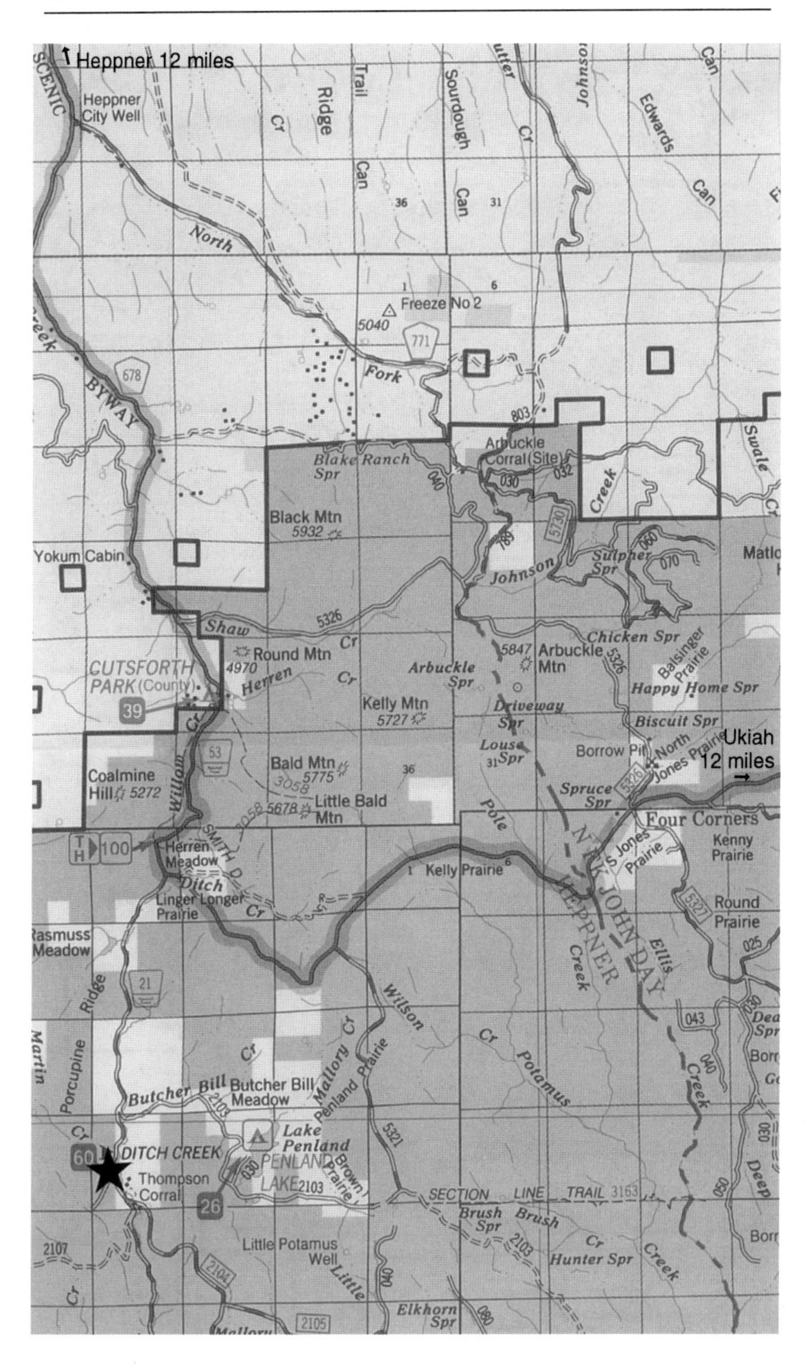

Ditch Creek Guard Station
Umatilla National Forest, Oregon

21 Clearwater Lookout Cabin

Umatilla National Forest

"The golden moments in the stream of life rush past us, and we see nothing but sand; the angels come to visit us, and we only know them when they are gone."

— GEORGE ELIOT

Your Bearings

25 miles south of Pomeroy, Washington
60 miles northeast of Walla Walla, Washington
90 miles west of Lewiston, Idaho
100 miles northeast of Pendleton, Oregon
150 miles south of Spokane, Washington

Availability Year-round.

Capacity Four people maximum. Ideal for families.

Description 14x21-foot, two-room cabin.

Cost $25 per night, plus a refundable $50 deposit.

Reservations

Available for as many as seven consecutive nights. For an application packet, maps, and further information contact:

Pomeroy Ranger District
Rt. 1, Box 53-F
Pomeroy, WA 99347
509-843-1891

How To Get There

Traveling through Pomeroy, Washington, on Highway 12 East (Main Street), you will notice toward the end of town a sign that reads UMATILLA NATIONAL FOREST 15 MILES. Follow this by turning right. This is 15th Street and becomes, in time, Road 128, Road 107, and Road 40—but this is all academic since most of these roads are not signed.

However, by following 15th Street south for 23 miles you will eventually reach the cabin. It is on the right. The road is paved to the National Forest boundary—about the first 15 miles. It is gravel thereafter but fairly well-maintained. The route is open to automobiles from about June 1 to November 1. Winter use may require travel by skis or snowshoes. Consult the District Office regarding current road and snow conditions prior to your departure.

Elevation 5600 feet.

Map Location Umatilla National Forest, Township 8 North, Range 42 East, Section 5.

What Is Provided

Propane cook stove, heater, light and fridge, a chest of drawers, desk, table, chairs and closet. The bedroom sleeps four, with two metal-framed single beds and a metal-framed double bunk.

The cabin is in need of the kind of loving attention that perhaps some local volunteers, in cooperation with the Forest Service, will give. The grounds are bereft of trees or shade, leaving the cabin itself exposed and looking somewhat bedraggled. It has only a very limited view and, despite its name, has no water, clear or otherwise.

What To Bring

All the water that you will need. Check with Ranger District regarding propane fuel supply.

History

The cabin was built in 1935 as a residence for the Clearwater Lookout guard. It still serves that purpose occasionally.

Around You

The trailhead for Bear Creek Trail (3110) is a mile south on Forest Road 40. On its way south to the Wenaha-Tucannon Wilderness in the Blue Mountains, it connects to Trail 3135, which follows the Tucannon River northwestward to a different part of the same wilderness.

Burley Mountain lookout guard.

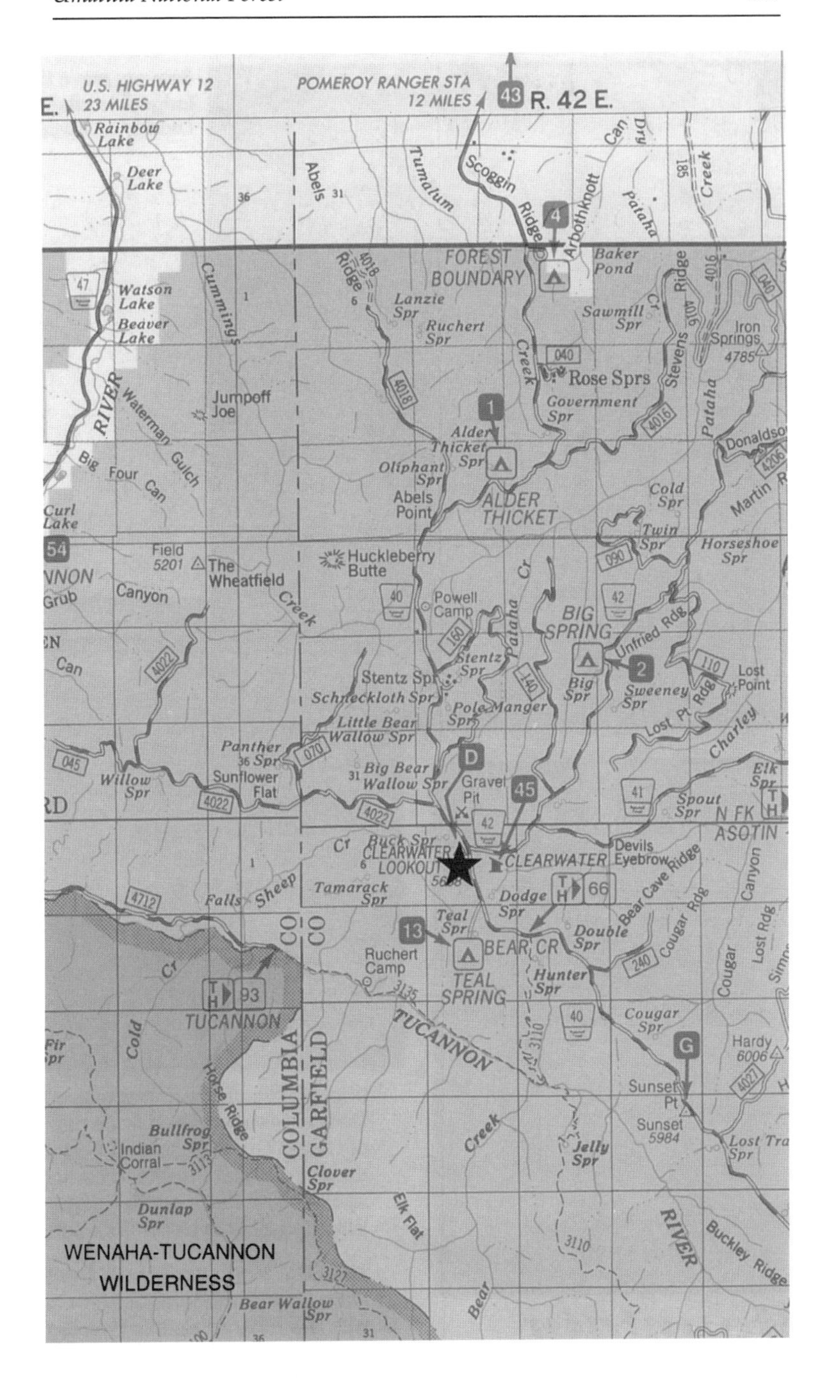

Clearwater Lookout Cabin
Umatilla National Forest, Washington

22 Godman Guard Station

Umatilla National Forest

"The bluebird carries the sky on his back."

— HENRY DAVID THOREAU, *Journal*, APRIL 3, 1852

Your Bearings

30 miles southeast of Dayton, Washington
60 miles east of Walla Walla
95 miles southwest of Lewiston, Idaho (via Dayton)
100 miles northeast of Pendleton
155 miles south of Spokane

Availability Year-round.

Capacity Ten people maximum, though half that number would be much more comfortable. Ideal for families.

Description

A nicely secluded 2-story guard station with living room, kitchen, bathroom and two bedrooms. There is a smaller house (with woodstove) adjacent to the guard station; it is available to skiers and snowshoers as a free day-use warming shelter.

Cost $25 to $60 per night, depending on the number of guests, plus a refundable $50 deposit.

Reservations

Available for as many as seven consecutive nights. For an application packet, maps, and further information contact:

Pomeroy Ranger District
Rt. 1, Box 53-F
Pomeroy, WA 99347
509-843-1891

How To Get There

The route is open to motor vehicles from about June 1 to November 1. Winter use may require travel by skis or snowshoes. Consult the District Office regarding current road and snow conditions prior to your departure.

Getting to this place can be as confusing as finding your way around the Internet. The Forest Service directions make getting there seem simple—except that not one of the county roads is signed, which renders those directions useless, as we found, having gotten lost several times. Here, whenever possible, we will use the road and street names that are signed, rather than guiding you by their map numbers.

From Main Street of Dayton, Washington—Highway 12 East—turn right on to 4th Street. Follow 4th Street for about 1/2-mile and turn left onto Eckler Street—*not* Eckler Mountain Road. Eckler Street becomes East Mustard Street, which leads to Skyline Drive. This is the road that goes to the National Forest and eventually becomes Forest Road 46, but not before you drive through about 18 miles of wheat fields. It is about 28 miles to the Godman Guard Station.

Just before the Godman Campground and a horse barn, turn left onto Forest Road 4608—the guard station is a few hundred yards up this road.

Elevation 5600 feet.

Map Location Umatilla National Forest, Township 7 North, Range 40 East, Section 10.

What Is Provided

Indoor plumbing—though in the winter the water is turned off to prevent pipe-freeze, so in winter you will need to bring your own drinking water, or have the means to treat the local supply. Check with the Ranger District.

The living room has propane lights, woodstove, couch, chest of drawers, a nice table, and a linoleum floor. The kitchen has a propane cook stove and fridge, and a sink. The downstairs bedroom has a wooden floor and three single metal-framed beds with less-than-inviting mattresses. The upstairs bedroom has two double beds and one single bed.

What To Bring

Drinking water or the means to treat the local supply during the winter months.

History

The Godman region has been used by area residents for summer camps and "tent homes" since the early 1900s. Construction of the guard station (the big house) began in 1933 when a Civilian Conservation Corps camp was located in the area.

Around You

The guard station is nicely secluded down in a hollow. Considering that there is a horse barn on the brow of the hill between the cabin and the wilderness, and a campground across the road from the horse barn, this seclusion is a blessing. A lovely creek runs through the back yard.

You are on the very rim of the Wenaha-Tucannon Wilderness portion of the Blue Mountains. The Godman Trail (3138) will lead you as deeply into it as you want to go.

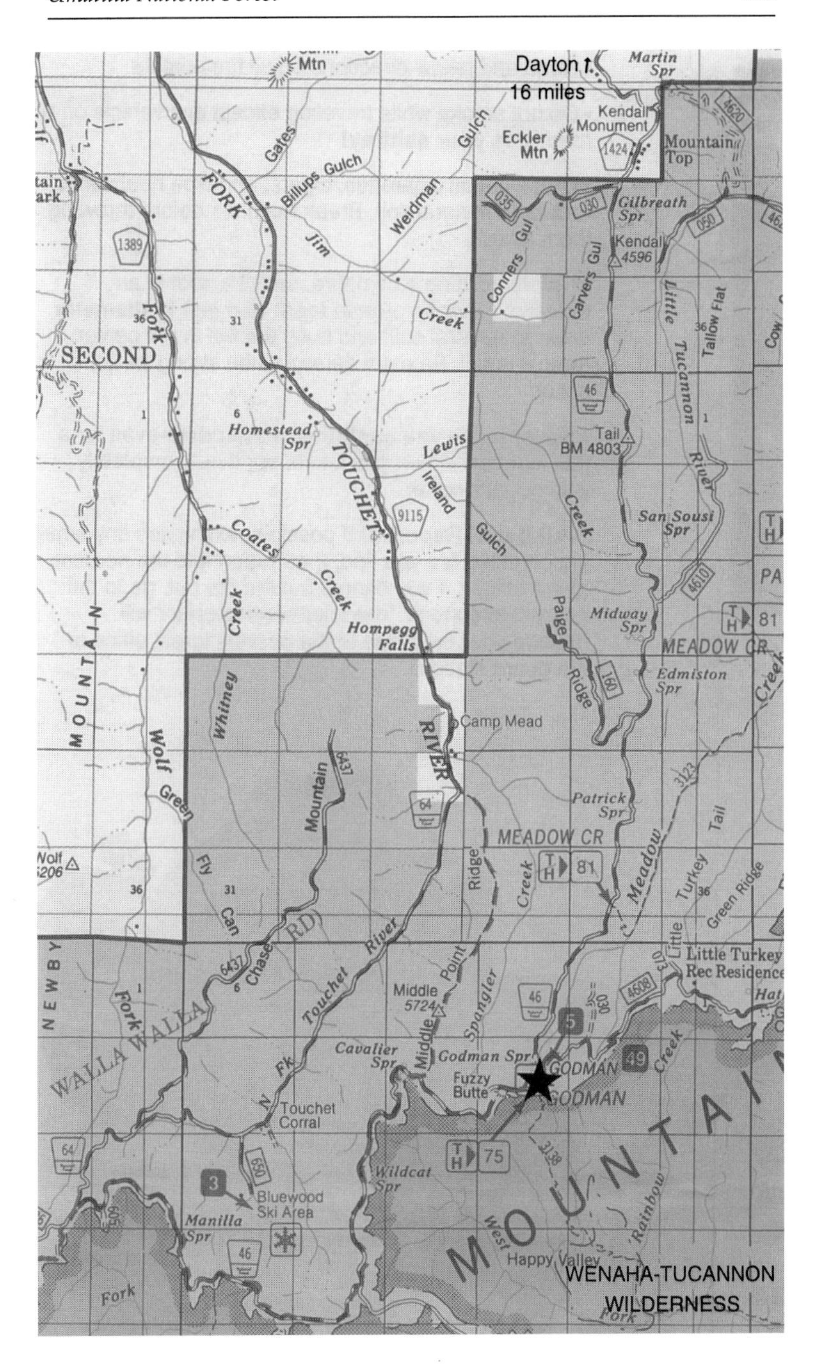

Godman Guard Station
Umatilla National Forest, Washington

23 Wenatchee Guard Station

Umatilla National Forest

"It appeared to be the most beautiful valley I had ever looked upon...thousands of ponies grazing, and Indians driving in all directions."

— JOHN JOHNSON, JULY 1851,
ON SEEING THE GRANDE RONDE VALLEY

Your Bearings

35 miles southeast of Pomeroy, Washington
70 miles southwest of Lewiston, Idaho (via Pomeroy)
105 miles northeast of Walla Walla
145 miles northeast of Pendleton
180 miles south of Spokane

Availability Year-round.

Capacity Four people maximum, though two would be more comfortable. Good for families.

Description One-story cabin with living room, bedroom and small kitchen. Secluded and comfortable, with extraordinary views.

Cost $25 per night, plus a refundable $50 deposit.

Reservations

Available for as many as seven consecutive nights. For an application packet, maps, and further information contact:

Pomeroy Ranger District
Rt. 1, Box 53-F
Pomeroy, WA 99347
509-843-1891

How To Get There

The route is open to motor vehicles from about June 1 to November 1. Winter use may require travel by skis or snowshoes. Consult the District Office regarding current road and snow conditions prior to your departure.

Traveling through Pomeroy, Washington, on Highway 12 East (Main Street), you will notice toward the end of town a sign that reads UMATILLA NATIONAL FOREST 15. Follow this by turning right. This is 15th Street which, over the next 15 miles, becomes, respectively, Roads 128, 107, and Forest Road 40 at the National Forest boundary. Here its surface changes to fairly well-maintained gravel. Continue on Road 40 another 16 miles to where it becomes Forest Road 44. After another 3-1/2 miles continue straight ahead—off Forest Road 44 and onto Forest Road 43. These roads are well signed. Go east on Road 43 for another three miles. The Guard Station is on a hill on the left side of the road. It is clearly visible.

But even before reaching the Guard Station you will be treated to astonishing vistas of Wenatchee Creek Canyon, Grande Ronde Valley, Wenaha-Tucannon Wilderness, and more.

Elevation 6050 feet.

Map Location Umatilla National Forest, Township 8 North, Range 43 East, Section 35.

What Is Provided

Propane heat, lights, cook stove and fridge. The small kitchen has a sink, but no running water. The bedroom has two single beds, and the living room/dining room also has a single bed.

What To Bring

Bring your own water, or the means to treat the local supply.

Setting

This historic guard station has a very cozy and comfortable ambience. It is beautifully proportioned and exquisitely secluded atop a ledge that looks into eternity.

History

Built in 1933 by Civilian Conservation Corps enrollees stationed in the District.

Around You

Beside the cabin, off Forest Road 43, is the trailhead for the Ranger Creek Trail (3137), which follows Ranger Creek to the Menatchee Creek, and follows that all the way to the Grande Ronde Valley.

The simplest way to reach Wenaha-Tucannon Wilderness is to go west from the cabin, on Forest Road 43, to Misery Spring Campground. At the campground turn left onto Forest Road 4030 and travel one mile to Kelly Camp Trailhead (3120).

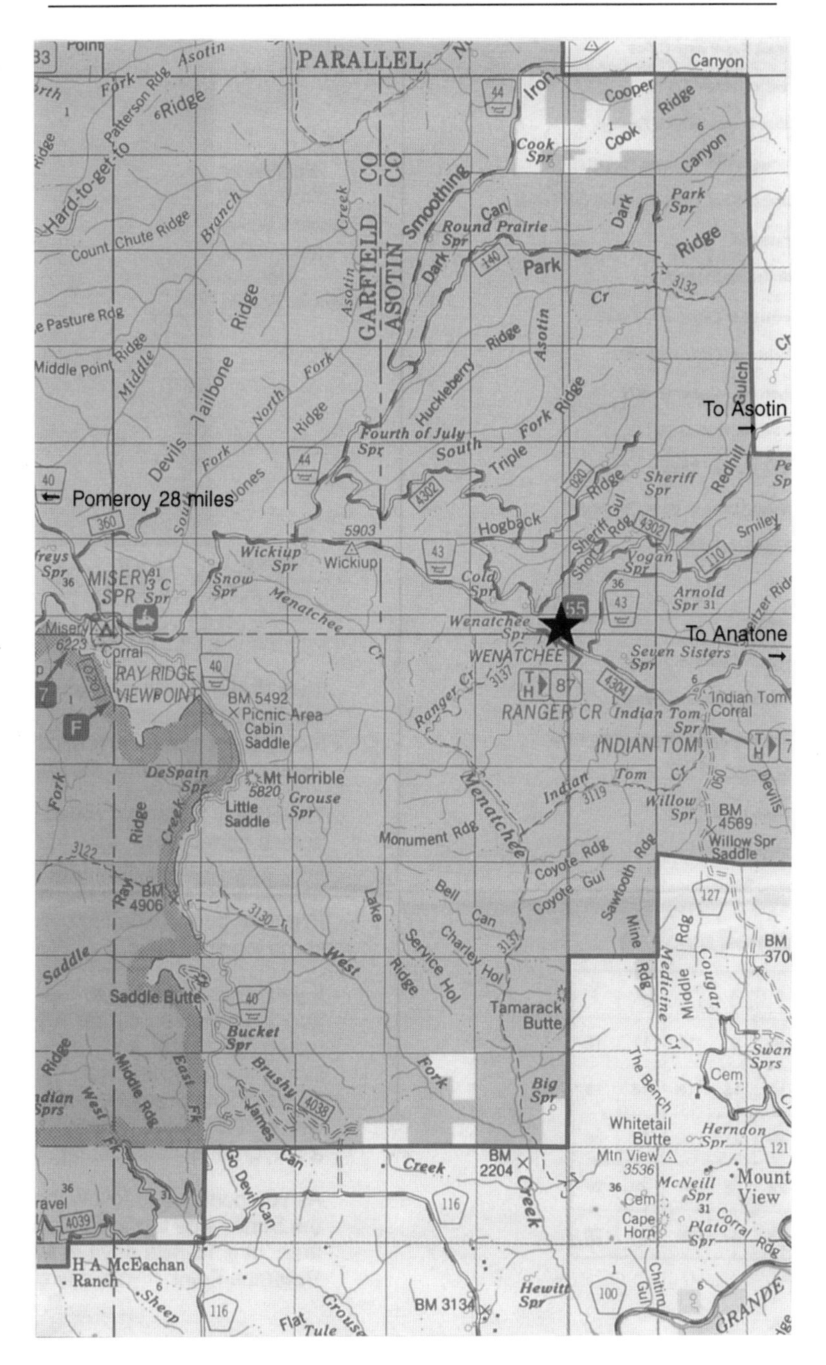

Wenatchee Guard Station

Umatilla National Forest, Washington

24 Pearson Meadows Guard Station

Umatilla National Forest

"Man is rich in proportion to the number of things which he can afford to let alone."

— Henry David Thoreau, *Walden*

Available 1996/97.
Contact the Ranger District for precise rental date.

Your Bearings

65 miles south of Pendleton
70 miles southwest of La Grande
75 miles northeast of Baker City
90 miles north of the town of John Day

Availability Year-round.

Capacity Four people maximum. Good for families.

Description

288 square feet, with bedroom, living room, kitchen, bathroom with shower.

Cost $25 per night.

Reservations

Available for as many as seven consecutive nights. For an application packet, maps, and further information contact:

North Fork John Day Ranger District
PO Box 158
Ukiah, OR 97880
541-427-3231

How To Get There

From Pendleton travel south on Highway 395 48 miles to State Highway 244. Turn left for Ukiah, which is about two miles east of 395. You are now on the Blue Mountain Scenic Byway which takes you all the way to Pearson Meadows Guard Station. In Ukiah turn right off Highway 244 onto Road 1475, which becomes Forest Road 52. Travel 14-1/2 miles south and southeast on Road 52 to the Guard Station. It will be on the right, on the south side of Road 52, and you will need to watch your odometer since it is not signed.

Four miles south of Ukiah is the Bridge Creek Wildlife Area—a wintering place for elk; nine miles farther east on the Byway is the North Fork John Day Overlook. On a clear day you should be able to see the Strawberry Mountains.

Elevation 5430 feet.

Map Location Umatilla National Forest, Township 6 South, Range 33 East, Section 21.

What Is Provided

Propane cook stove and fuel, fridge, lights, tables and chairs, and running water (non-potable).

What To Bring

Drinking water, or the means to treat the local supply.

Setting

The cabin is quaint but a bit distressed and in need of the kind of comforting that only the Forest Service can provide. However, if you can accept its shortcomings, it is pleasantly situated overlooking a lovely meadow, yet very close and convenient to the Byway, and it is surrounded by huge old trees.

History

Constructed in the mid-1930s during the Great Depression by the Civilian Conservation Corps as a Forest Service administrative site. The guard station is now used periodically by Forest Service fire crews.

Around You

There are several trailheads off the Scenic Byway, Road 52, close to the cabin. Approximately eight miles east of the cabin, Big Creek Trail (3166) goes into the section of North Fork John Day Wilderness that is south of the Byway. Trail (3169) goes into the section of North Fork John Day Wilderness that is north of the Byway.

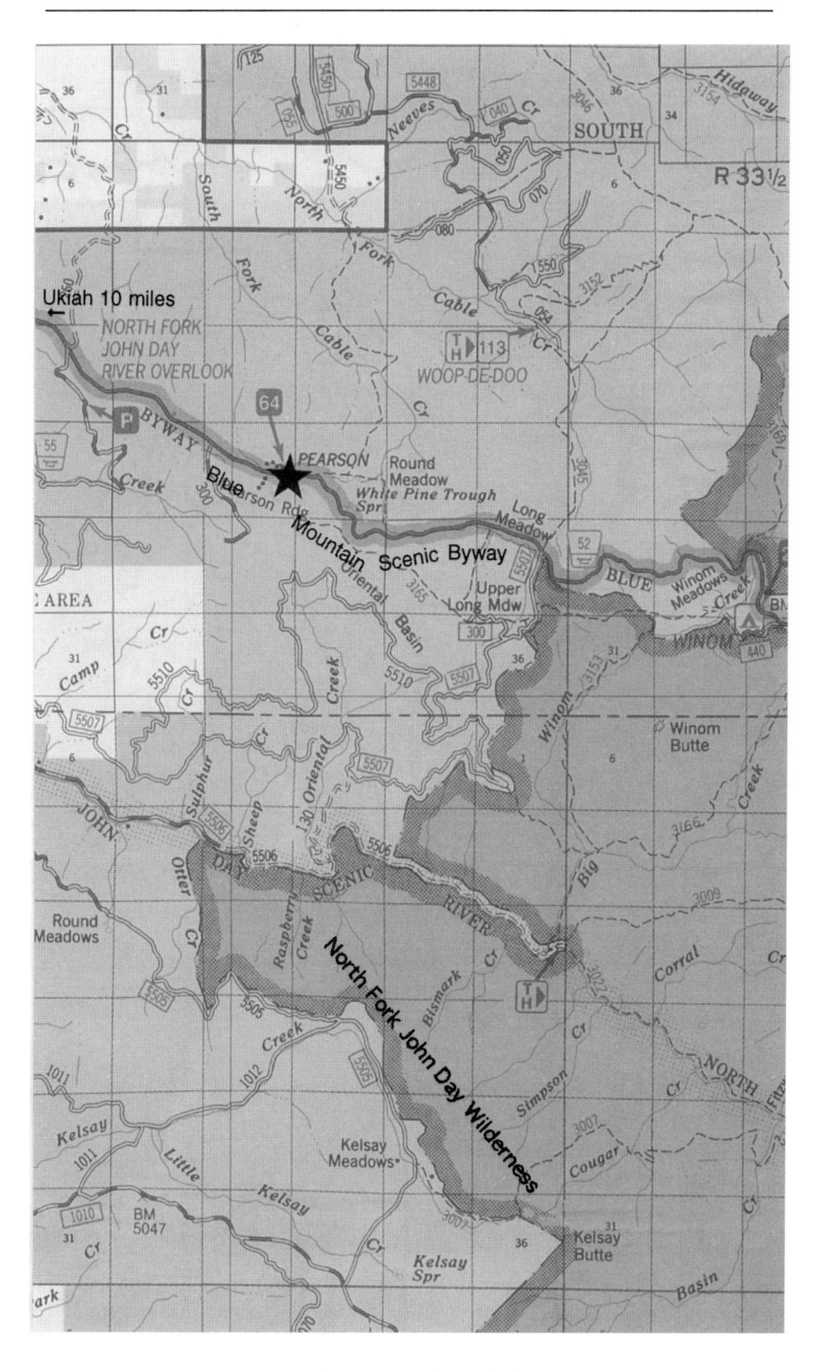

Pearson Meadows Guard Station
Umatilla National Forest, Oregon

25 Fry Meadow Guard Station

Umatilla National Forest

"Life consists with wilderness. The most alive is the wildest. Not yet subdued to man, its presence refreshes him."

— HENRY DAVID THOREAU, 1862

Available 1997/98.
Contact the Ranger District for precise rental date.

Your Bearings

50 miles northeast of La Grande
80 miles northwest of Enterprise
90 miles north of Baker City
100 miles east of Pendleton

Availability Year-round.

Capacity Four people maximum. Ideal for families.

Description Small cabin with bedroom, living room and kitchen.

Cost $25 per night, plus a refundable $50 deposit.

Reservations

Available for as many as seven consecutive nights. For an application packet, maps, and further information contact:

Walla Walla Ranger District
1415 West Rose
Walla Walla, WA 99362
509-522-6290

How To Get There

The route is open to vehicular travel from about May 1 to December 1. Winter use may require you to travel by skis or snowshoes. Consult the District Office regarding current road and snow conditions prior to your departure.

The Forest Service directions make getting there seem simple—except that most of the roads are not signed, which renders those directions useless, as we found, having gotten lost several times. Here, whenever possible, we will use the road and street names that are signed, rather than guiding you by their map numbers.

From La Grande, take Highway 82 northeast 20 miles to the town of Elgin. This is where it gets sticky. At the main crossroads in the town of Elgin turn left. After about one mile, turn right. Follow the signs for LOOKINGGLASS FISH HATCHERY and PALMER JUNCTION via Middle Road and Gordon Creek Road for about 20 miles, to Palmer Junction. These roads are numbered on the map as County Road 42, but are not signed as such.

Continue just a few hundred yards past Palmer Junction on Road 42 (which is signed Moses Creek Lane—do not take Lookingglass Road) until you come to Lookout Mountain Road (Forest Road 43). Turn left here and continue on this poorly maintained, winding, steep road for eight miles—it becomes Forest Road 6231. You will finally see a sign for Fry Meadow. Turn left at this sign; this is Forest Road 6235. The cabin is on your left about 100 yards after the turn, set at the sylvan corner of a lovely meadow.

A longer route, but with a slightly better road, is to follow the BOWMAN LOOP and JUBILEE LAKE signs from Road 42 two miles west of Palmer Junction. Go north on Forest Road 63 (unmarked) for five miles to Forest Road 62. Turn right (northeast) onto Road 62, and continue four miles to Forest Road 6235. Turn right onto Road 6235. The cabin is a mile ahead on the right, 100 yards from the junction of Road 6231.

Elevation 4138 feet.

Map Location Umatilla National Forest, Township 4 North, Range 40 East, Section 32.

What Is Provided

Table and chairs, a stove, and two sets of bunk beds which sleep four. Outside, a picnic table, firering and vault toilet.

What To Bring

Drinking water; check with Ranger District regarding local, untreated supply.

Setting

The cabin, surrounded by fine old trees, peacefully overlooks a meadow that was luscious with wildflowers the day we were there; and that evening, as we sat on the picnic table by the firering when the birds began their evening song with full-throated ease, the meadow became Keats' "melodious plot of beechen green, and shadows numberless."

History

Fry Meadow Guard Station is south of Wenaha-Tucannon Wilderness and west of the Wild and Scenic Grande Ronde River. The Grande Ronde Valley was named by a French fur-trapper for its circular shape. From 1843 on, hundreds of thousands of emigrants passed through here on their way to the Willamette Valley.

In better days—for Native Americans, anyway—the Nez Perce, Cayuse, Umatilla, Bannock, Yakima, and Walla Walla tribes shared the Grande Ronde Valley and its hot springs and food in all its natural abundance.

John C. Fremont, when he saw this valley in October 1843, wrote in his journal, *"It is a place—one of the few we have seen in our journey so far—where a farmer would delight to establish himself, if he were content to live in the seclusion it imposes...it may in time, form a superb country."*

Fry Meadow was first established as a Forest Service Ranger Station sometime before 1908 for what was then called Wenaha National Forest. The Guard Station was built in the early 1930s by the Civilian Conservation Corps; it is a classic example of their Depression Era architectural styling and construction.

Lookingglass Fish Hatchery gets its interesting name from a Nez Perce leader called Apash-wa-hay-ikt, but called Chief Looking Glass by Europeans because he often carried a hand-mirror with him.

Around You

The 200 miles of trails in Wenaha-Tucannon Wilderness can be reached by following either Forest Roads 6231 north or 6235 northwest to Forest Road 62. Go north on Road 62 to its junction with Forest Road 6413. Turn left onto Road 6413, and then right onto Road 6415 to Trail 3236 and the wilderness entry. In addition, the entire area around Fry Meadow is a maze of logging roads. All invite a saunter.

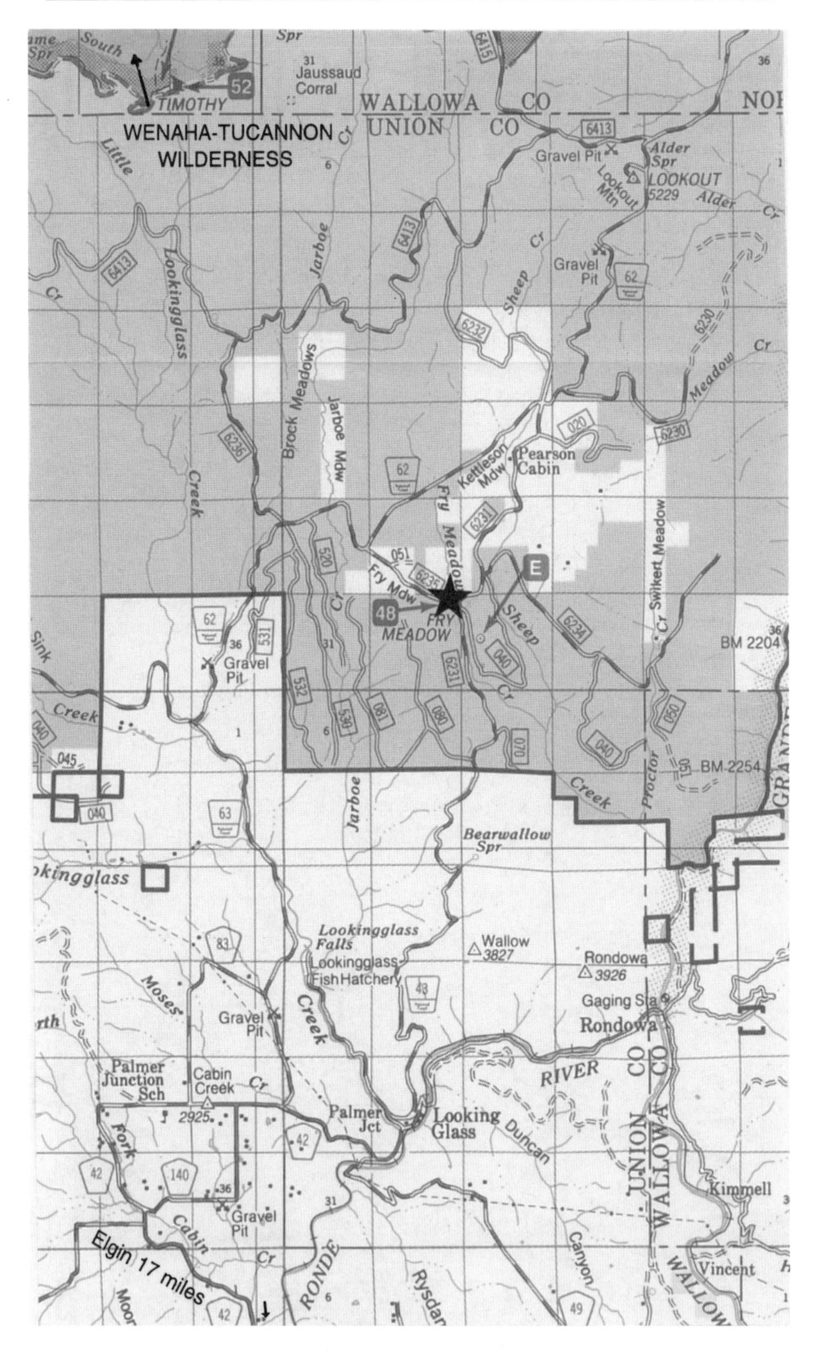

Fry Meadow Guard Station
Umatilla National Forest, Oregon

26 Summit Guard Station Bunkhouse
Umatilla National Forest

"I frequently met old acquaintances, in the trees and flowers, and was not a little delighted. Indeed I do not know as I was ever so much affected with any scenery in my life . . ."

— From the diary of Narcissa Whitman, pioneer to this area, 1836

Available 1996/97.
Contact the Ranger District for precise rental date.

Your Bearings

35 miles north of La Grande
50 miles southeast of Pendleton
75 miles north of Baker City
90 miles south of Walla Walla

Availability Year-round.

Capacity Four people maximum. Good for families.

Description

24x15-foot one-room bunkhouse with a very low ceiling. In need of some Forest Service T.L.C., but with panoramic views.

Cost $25 per night, plus a refundable $50 deposit.

Reservations

Available for as many as seven consecutive nights. For an application packet, maps, and further information contact:

Walla Walla Ranger District
1415 West Rose
Walla Walla, WA 99362
509-522-6290

How To Get There

The route is open to vehicular travel from about May 1 to December 1. Winter use may require you to travel by skis or snowshoes. Consult the District Office regarding current road and snow conditions prior to your departure.

From Pendleton travel southeast on Interstate 84 for about 38 miles and take the SUMMIT ROAD—MT. EMILY exit. Go northeast on Summit Road (Forest Road 31), which is gravel, for 12 miles to its junction with Forest Road 3113. Turn left here; the bunkhouse is a mile ahead on the left. Note that Summit Guard Station, the first structure on the left, is not for rent. The Bunkhouse is the only cabin here available for overnight guests.

Elevation 4780 feet.

Map Location Umatilla National Forest, Township 1 South, Range 37 East, Section 17.

What Is Provided

Four single beds, a cook stove, table, chairs, sink and closets, a linoleum floor and a vault toilet.

What To Bring

Drinking water or the means to treat the local supply.

History

The Summit Ranger Station was established here sometime before 1908 as an administrative site for what used to be called Wenaha National Forest. The original cabin was replaced with a residence built by the Civilian Conservation Corps in 1934. The Guard Station is still used by Forest Service fire crews. The bunkhouse itself was built in the 1970s and is purely utilitarian, entirely lacking the charm and grace of the work done in earlier times, but, from its position atop Drumhill Ridge, it does preside over a grand and expansive view.

Around You

Magnificent views overlooking sections of the Grande Ronde Valley, with snow-capped peaks in the distance. On the way up Forest Road 31 you will find an interpretive sign a short distance before Forest Road 3109, showing where Marcus and Narcissa Whitman, with Reverend and Eliza Spalding, crossed the summit in 1836. Ms. Whitman and Ms. Spalding became the first European women to cross the continent traveling this route.

On graveled Forest Road 3109 stop at the Whitman Route Overlook for a sweeping vista of this rugged and remote terrain. The bunkhouse itself overlooks a chunk of the Blue Mountains that includes Sugarloaf, Spring, Green and Wilbur mountains. In addition, it is surrounded by massive old-growth trees that provide day-long shade. There were deer prancing around much of the afternoon we spent there, and not a single car went by nor did we encounter anyone on our way to or from the cabin; though a coyote did eye us knowingly on the way back.

We have added the Bear Creek Trail to the map. The trailhead is within walking distance of the cabin, and travels through old-growth forest to the North Fork Meacham Creek, which it follows for 2-1/2 miles; then climbs 3-1/2 miles to the top of Thimbleberry Mountain.

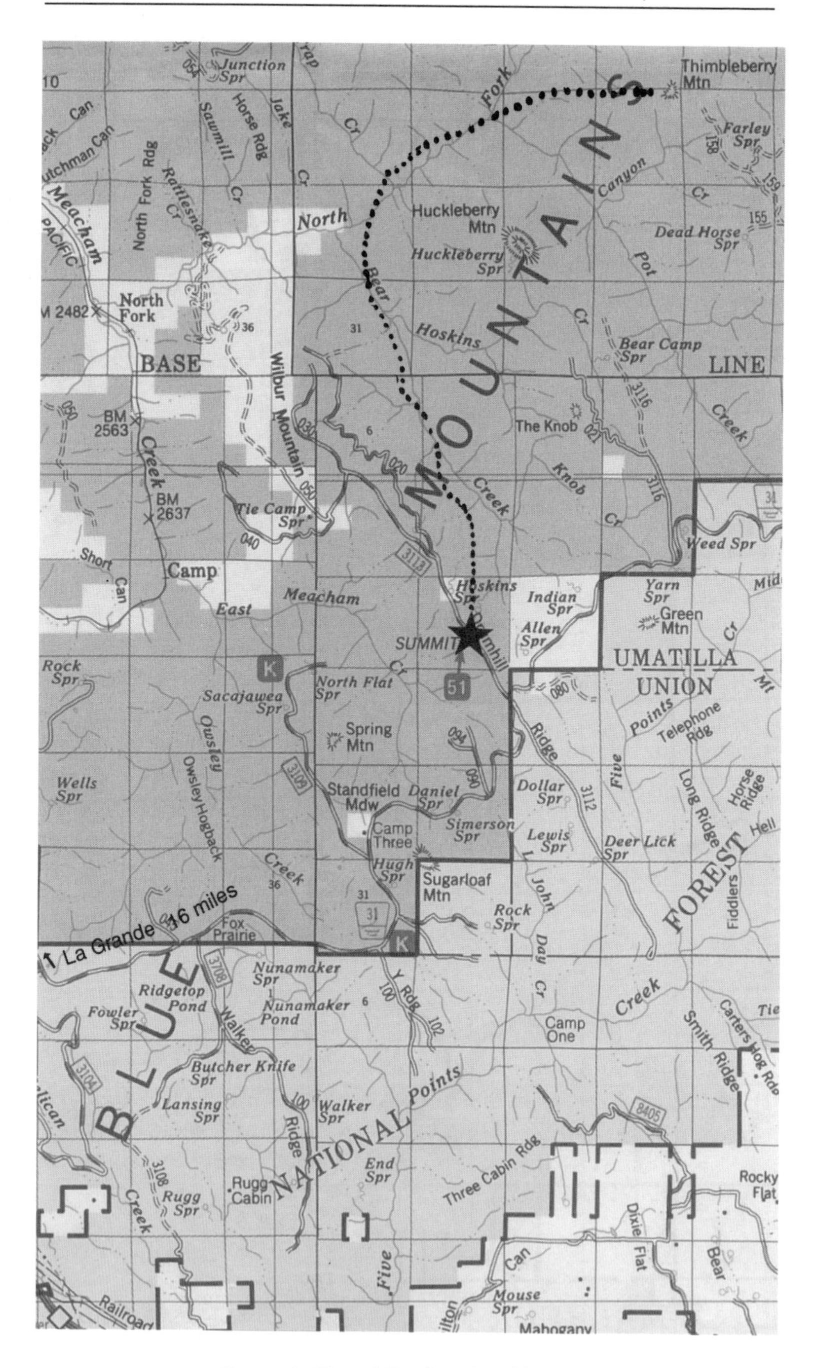

Summit Guard Station Bunkhouse
Umatilla National Forest, Oregon

27 Two Color Guard Station

Wallowa-Whitman National Forest

"And to glance with an eye or show a bean in its pod confounds the learning of all times . . ."

— Walt Whitman, *"Song of Myself"*

Your Bearings

45 miles northeast of Baker city
55 miles southeast of La Grande
105 miles southeast of Pendleton

Availability Year-round.

Capacity

As many as 12 people, though half that number would be more comfortable. Ideal for families. Pets okay if kept outside and leashed.

Description

One-story cabin with over 1000 square feet of floor space and a lovely back porch overlooking the Wild and Scenic Eagle Creek.

Cost

From one to four people the cost is $40 per night. The fee increases $5 a night for each additional person up to 12. Refundable $50 deposit. You must send in your rental fees, including the deposit, at least three weeks prior to your confirmed rental dates.

Reservations

In the spring, summer, and fall this is a popular get-away, so make your reservations early. Available for as many as seven consecutive nights. For an application packet, maps, and further information contact:

La Grande Ranger District
3502 Highway 30
La Grande, OR 97850
541-963-7186

How To Get There

For much of the winter the final 16 miles into the cabin are navigable only by snowmobile. The guard station is located at an elevation that gets sudden and heavy snowfall. Even with a small accumulation the access road can be dangerous. In winter, especially, keep posted on the weather reports and contact the Ranger Station for current conditions and advisories just prior to your trip.

There are no signs along the way for Two Color Guard Station. From Baker City, Oregon, travel six miles north on Interstate 84. Exit east at the sign for MEDICAL SPRINGS onto Highway 203, and follow it for 20 miles to Medical Springs. There, turn right, following the sign for BOULDER PARK. This is Big Creek Road, which becomes Forest Road 67.

After 17 miles beyond Medical Springs—17 long, winding, contorted, washboarded, gravel miles with extraordinarily lovely views of surrounding mountain peaks still covered in snow, even in July—you will be glad to reach Forest Road 77. Turn left.

You are now beside roaring, foaming, exquisite Eagle Creek. After less than a mile, Forest Road 77 veers off to the left, but you continue straight on Forest Road 7755, also called Two Color Road. After 1-1/2 miles be on the lookout for Two Color

Guard Station. It is on the right and easy to miss—but well past the Two Color Campground.

Elevation 4825 feet.

Map Location Wallowa-Whitman National Forest, Township 6 South, Range 43 East, Section 15.

What Is Provided

Ten single beds, one double bed, woodstove for heat, propane cook stove and lights, cups, place settings for 12, eating utensils for 12, potable water, wash basins, firewood, ax, shovel, even a shower.

Outside you will find a spacious yard with two picnic tables, firering, a corral large enough for several horses, and a watering trough.

Firewood is in the woodshed. The house key will open its lock. A pipe 15 feet east of the outhouse has spring water running from it. During the winter months, the inside toilet can be used by filling the cistern with water before flushing—but do not leave any water in there when you leave. Put small amounts of anti-freeze in all drains and toilets—you will find the anti-freeze under the kitchen sink.

What To Bring

From October to May the potable water is shut off to prevent pipe-freeze, therefore inquire at the Ranger District whether you need to bring drinking water.

History

Built in 1959 as a backcountry residence for field crews working on the La Grande Ranger District.

Around You

Tucked away in a secluded corner of Wallowa-Whitman National Forest in northeastern Oregon, the cabin is just a two-hour drive from civilization. It overlooks a riparian marsh along the banks of Eagle Creek, which is noted for its rainbow trout.

In early July, when we were here, especially in the late evening, what was even louder than the gurgling of Eagle Creek was the offensive laughter of mosquitoes with stingers sharp enough to penetrate a coat of medieval armor. Maybe we had been in the woods too long, but it seemed at the time they were laughing at our mosquito repellent.

If steep and rugged hikes to mountain lakes interest you, Eagle Cap Wilderness, which has no less than 18 mountains over 9000 feet, is just two miles northeast at the end of Forest Road 7755. Here you will find Boulder Park Trail 1922, which connects to Trail 1921—for Culver Lake (5.2 miles), Bear Lake (3.8 miles), and Looking Glass Lake (6.2 miles); and Trail 1931 for Eagle Lake and other glorious destinations beyond.

Three Sisters Mountains from Indian Ridge lookout.

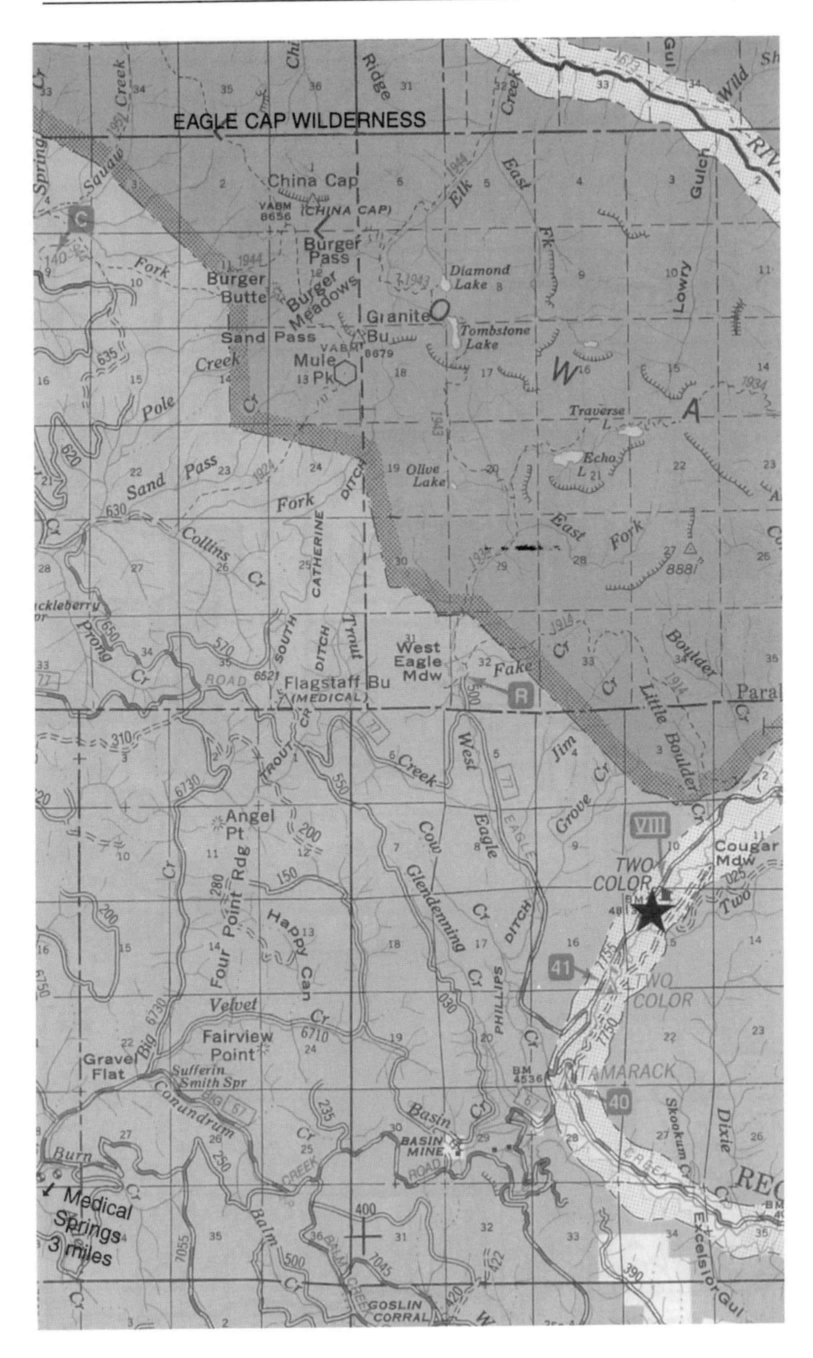

Two Color Guard Station
Wallowa-Whitman National Forest, Oregon

28 Antlers Guard Station

Wallowa-Whitman National Forest

"It is a great art to saunter."

— Henry David Thoreau, *Journal*, April 26, 1841

Your Bearings

35 miles northeast of Prairie City
40 miles west of Baker City
80 miles southwest of La Grande
130 miles south of Pendleton

Availability Year-round.

Capacity

Maximum group size is six, though there are enough beds to sleep only four. Ideal for families. Pets okay if kept outside and leashed.

Description

Small two-room cabin with fenced yard. Very pleasant and will maintained.

Cost $25 per night, plus a refundable $50 deposit.

Reservations

Available for as many as seven consecutive nights. Check in is at 12:00 noon and check out is at 10:00 AM. For an application packet, maps, and further information contact:

Unity Ranger District
214 Main Street
PO Box 36
Unity, OR 97884
541-446-3351

How To Get There

From Prairie City, Oregon, travel northeast on Highway 26 for 16 miles; turn left on Highway 7 and follow it north and northeast for 17 miles to County Road 529. Turn right on Road 529. This gravel road goes through the abandoned town of Whitney. After 2-1/2 miles on Road 529 you will see Antlers Guard Station on the left, tucked away under a steep hillside.

From Baker City, Oregon, travel south and then west on Highway 7 (part of the Elkhorn Scenic Byway), past 2400-acre Phillips Reservoir. Remain on Highway 7 until you reach County Road 529, about 38 miles. Turn left (south) onto Road 529, and travel through the abandoned Whitney town site. Antlers Guard Station is about 2-1/2 miles south on this gravel road, on the left.

From Unity, Oregon, travel north on Highway 26 approximately two miles, to Highway 245. Turn right. Continue on Highway 245 for approximately six miles to County Road 535. Turn left and continue northwest on Road 535—which becomes Road 529—and meanders alongside the North Fork of the Burnt River on its way to Antlers Guard Station, which is on the right side of the road.

During the winter months, Roads 529 and 535 are plowed intermittently. Contact the Ranger District in Unity for current road conditions.

Elevation 4107 feet.

Map Location Wallowa-Whitman National Forest, Township 11 South, Range 36 East, Section 3.

What Is Provided

Eating utensils for six, a woodstove, potable water from a hand-pumped artesian well, enough beds to sleep four, a propane cook stove, propane lights, ax and shovel, water bucket, an outhouse, picnic table and firering.

On the day we were there the cabin was being rented by a family—two adults and three children—who were enjoying a barbecue in the front yard. They spoke fondly of the cabin and of their time there.

What To Bring

Your family and friends.

History

Built by the Civilian Conservation Corps in the 1930s and used as the field headquarters for fire patrols.

Around You

It is legal to pan for gold in this section of the Burnt River, which flows through the backyard. Although we found no holes in the immediate vicinity big enough to actually swim in, there were several spots in the river suitable for wading.

The townsite of Whitney is about two miles north on Road 529. It once had a population of 150, but the entire town closed down when the trains stopped running through this area.

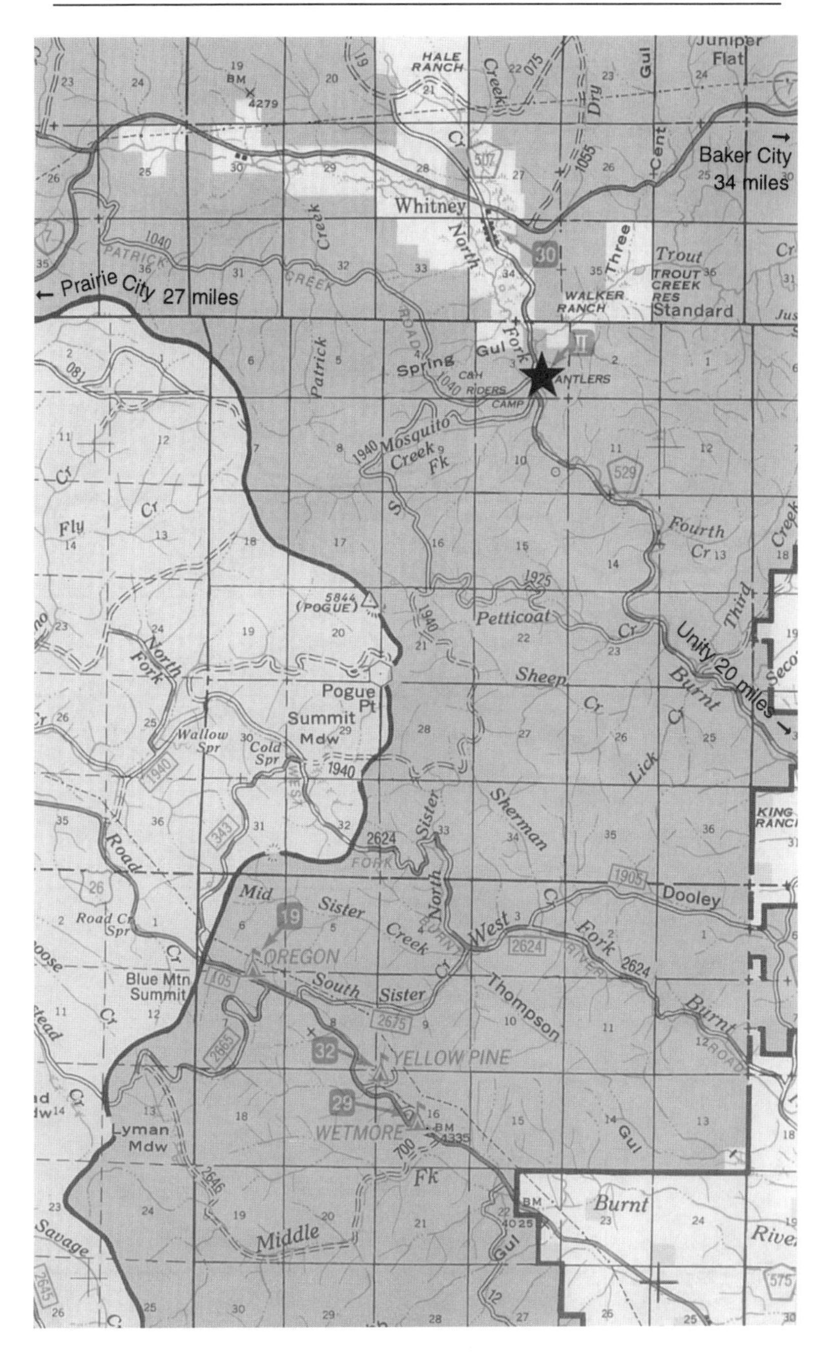

Antlers Guard Station
Wallowa-Whitman National Forest, Oregon

29 Peavy Cabin

Wallowa-Whitman National Forest

"Time is but a stream I go a-fishing in."

— HENRY DAVID THOREAU

Your Bearings

50 miles northwest of Baker City
60 miles southwest of La Grande
90 miles south of Pendleton

Availability Year-round.

Capacity Four people. Ideal for families.

Description

Historic, one-room log cabin, measuring 16x24-feet. Intelligently restored to its original loveliness, almost to museum level. Beautiful and secluded.

Cost $40 per night, plus a refundable $50 deposit.

Reservations

Historic Peavy Cabin, a king of cabins, is a new arrival to the rental program. It has been completely refurbished, and is available for periods of up to 10 days. For an application packet, maps, and further information contact:

Baker Ranger District
3165 Tenth Street
Baker City, OR 97814
541-523-1932

How to Get There

From Baker City, Oregon, take Highway 30 north to Haines. Haines was known in the good old days as the town with "whiskey in the water and gold in the street." From Haines, turn west onto County Road 1146 and follow the ELKHORN DRIVE SCENIC BYWAY signs to Anthony Lakes, 34.4 miles. Continue for another 12.5 miles past the ski resort, past the Elkhorn Summit, and down to the North Fork John Day River and Forest Road 380, which is on the left. By now you will have traveled about 47 miles from Baker City. At the junction of Forest Road 380 there is a sign: PEAVY CABIN 3, CUNNINGHAM COVE 3, TRAIL 1643.

The road, paved all the way, passes through some of the grandest scenery to be found anywhere. Elkhorn Summit, at 7392 feet, is the highest ground to be reached by paved road in all of Oregon. The snow-covered peaks of the Elkhorn seem close enough to reach out and touch.

When you turn left onto Forest Road 380 you will see a sign that reads ROUGH ROAD AHEAD. NOT RECOMMENDED FOR PASSENGER CARS. Pay heed to this sign—park your car and walk. We made it all the way driving a 1971 Datsun pick-up, but only by getting out every now and then, especially during the last mile, to measure with a stick the depth of the water in the pools that completely hid the road—if one can call it that. Anyway, Thoreau would definitely want you to walk the rest of the way. The road follows the wonderful North Fork John Day River all the way to the cabin.

In winter, Forest Road 73 is not plowed beyond the Anthony Lakes Ski Area. Contact Baker Ranger District for current weather and road conditions

Elevation 5800 feet.

Map Location Wallowa-Whitman National Forest, Township 7 South, Range 36 East, Section 35.

What Is Provided

Woodstove, firewood, propane lamps, cook stove and fridge, fireplace, large table, benches, four single beds, eating utensils for four, cups, glasses, tea and coffee pots, ax and shovel, pots, pans and a single sink. Also vault toilet, tack shed, and partially fenced pasture.

What To Bring

Drinking water, or the means to treat local water.

Setting

If we were to be given the frightening task of finding a cabin to rent for no less an expert than Henry David Thoreau, we would, we are confident, after consulting his friend Emerson, choose Peavy Cabin. No other cabin in this book comes as close to Thoreau's most exacting criteria for a place to live.

One of the many reasons Peavy Cabin is so close to Thoreau's ideal is that it is situated on the edge of a wilderness, the North Fork John Day; another reason is that it is on the bank of a river, the North Fork of the John Day; and yet another is that a tiny stream flows by the house—through the front yard in fact. And, being the sturdy soul Thoreau undoubtedly was, there is Lost Lake, to which he could hike—or saunter, as he would prefer.

And, perhaps, even more important to Thoreau than any of these, would be that the cabin's location allows for solitude and seclusion. This, too, Peavy does, in great and generous measure, though this makes it quite difficult to get to.

It is set in a large and level meadow; on the day we were there in mid-July, there was a lovely stream of the clearest water flowing by. The cabin itself is exquisite in almost every detail, with log and stone blending beautifully together. The floors are of polished wood, the ceiling beams are unhewn, the fireplace is of solid stone, the woodstove is waist-high, the chairs and table seem especially made and designed for this cabin, and even the pots and pans hanging on the wall seem to belong to another, sturdier era.

History

Built in 1934 by Dr. George Wilcox Peavy, Dean of the School of Forestry at Oregon Agricultural College (later Oregon State University), who used it as an office for his field laboratory work. Its integrity and charm have survived several restoration projects, and it is now recorded in the Heritage Resource Inventory as an historic site.

Around You

Secluded on the edge of wilderness, Peavy Cabin stands between a dense alpine forest and an open meadow.

The Wild and Scenic North Fork of the John Day River is at your back door. At your front door is North Fork John Day Wilderness, and the trailhead for Cunningham Cove Trail (1643) is practically in the front yard. This trail takes you into the Wilderness and, after 3.3 miles, reaches the Elkhorn Crest Trail (1611), the highest in the Blue Mountains. It leads 24 miles from the Anthony Lakes area in the north to Marble Creek Pass in the south, reaching elevations of 8200 feet.

The Peavy Trail (1640) follows the North Fork John Day River and eventually leads into North Fork John Day Wilderness and reaches the Elkhorn Crest Trail (1611). This makes possible a difficult but spectacular 11-1/2 mile loop—by traveling north on the Elkhorn Crest Trail and returning to the cabin on the Cunningham Cove Trail (1643). There are tens of thousands of elk and mule deer in this wilderness.

"Shall I not have intelligence with the earth?
Am I not partly leaves and vegetable mould myself?"

— Henry David Thoreau

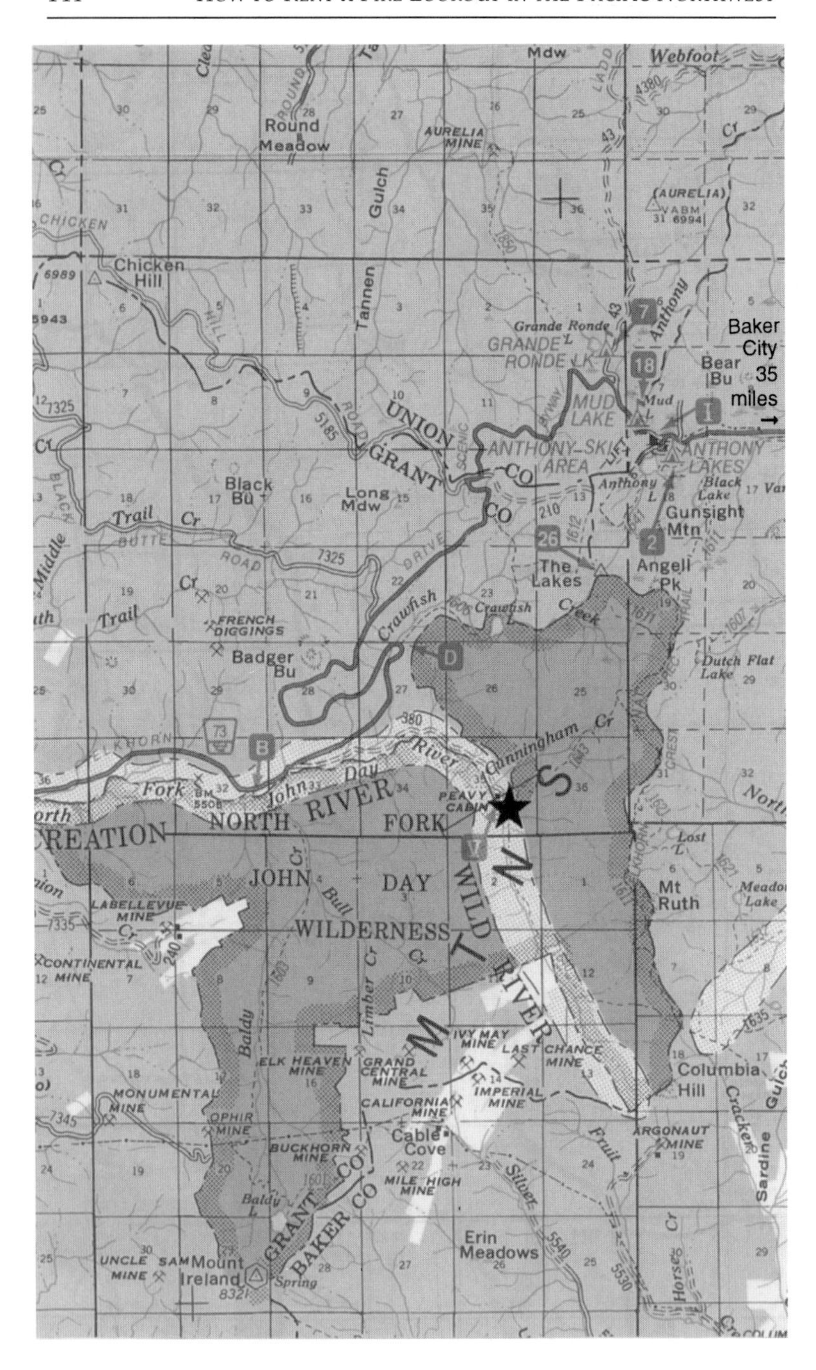

Peavy Cabin
Wallowa-Whitman National Forest, Oregon

30 Lily White Guard Station

Wallowa-Whitman National Forest

"Love is a call to action."

— Kevin Peer, Documentary Filmmaker

Your Bearings

40 miles northeast of Baker City
75 miles southeast of La Grande
125 miles southeast of Pendleton

Availability Year-round.

Capacity

As many as 15, though fewer would be more comfortable. Pets are welcome if they are kept outdoors and leashed.

Description

36x18-foot cabin with kitchen, dining room, and living room. Wheelchair-accessible.

Cost

$40 per night for four people with a $5 charge for each additional person up to a maximum of $95. Refundable $50 deposit required.

Reservations

Permits are issued for as many as 14 consecutive nights. Check-in and check-out time is 10:00 AM. For an application packet, maps, and further information contact:

Pine Ranger District
General Delivery
Halfway, OR 97834
541-742-7511

How To Get There

From Baker City, travel one mile north on Interstate 84. At Exit 302 take Highway 86 east for about 22 miles to Road 852, also called Sparta Grade Road. This is easy to pass by since the sign, SPARTA, is not very obvious. Sparta was one of the first mining centers in this region. Turn left (north). This is a steep, rough and winding, gravel road, but it does have magnificent views along the way.

After five miles on Road 852, turn left (north) onto Road 891. The sign reads: LILY WHITE. Follow this road for six miles (it becomes Forest Road 70) until you reach Forest Road 7020. Veer right onto Road 7020. Proceed 1-1/2 miles, during which you can catch glimpses of Lily White Guard Station through the trees on the left. Turn left onto Forest Road 160. The Guard Station is a couple of hundred yards away.

An alternate route is to travel south from Pine (not shown on map), on State Highway 86 to Richland, where you turn north to New Bridge and the Sparta Grade Road, Road 852. Go west on this gravel road, past the mining ghost town of Sparta, 12.8 miles, to Road 891. Turn right (north) onto Road 891 (which becomes Forest Road 70) and continue north for six miles until you reach Forest Road 7020. Veer right onto Road 7020. Proceed for about 1-1/2 miles, during which you can catch glimpses of Lily White Guard Station through the trees on the left. Turn left onto Forest Road 160. The Guard Station is a couple of hundred yards away.

During winter, Forest Road 70 is rarely plowed. Getting to Lily White then becomes an over-the-snow endeavor, via skis or snowshoes, for as many as seven miles, depending on snow conditions.

Elevation 4500 feet.

Map Location Wallowa-Whitman National Forest, Township 7 South, Range 44 East, Section 18.

What Is Provided

15 metal cots, though there are place settings and eating utensils for only six. Propane cook stove and lights, woodstove, firewood, potable water from an outdoor faucet in summer only, ax and shovel.

The living room and dining room can be separated by a sliding partition. The living room is carpeted, has a wood stove, two arm chairs, coffee and card tables. The dining room has a formica-top dining table and folding chairs.

The kitchen has a stove, sinks, lots of cupboards and countertop space. Though the indoor faucets do have potable running water, the Forest Service cannot allow them to be used until a grey-water system has been installed to disperse the used water. Meanwhile, there is a faucet outside the door with potable water. In the winter when that water is turned off, use the spring beside the house—which is where the house water comes from.

Though the Forest Service has made a good start revamping this cluster of buildings, there is still some work to be done.

What To Bring

Water in the winter or a means to treat the local supply.

History

There is a cluster of buildings here situated on 10 fenced acres. Only the main lodge, which was originally a mess hall is available to rent, together with the two wheelchair-accessible vault toilets.

Around You

The glorious Wild and Scenic Eagle Creek is a few miles east on the winding and washboarded Empire Gulch Road, Forest Road 7015.

At East Eagle Mine, four miles north on East Eagle Road, Forest Road 7745, which spurs off east from Forest Road 77, is the start of the Sullivan Trail (1946), which leads into Eagle Cap Wilderness and continues up to the Cliff River Trail (1885), within the wilderness. (Not shown on our map).

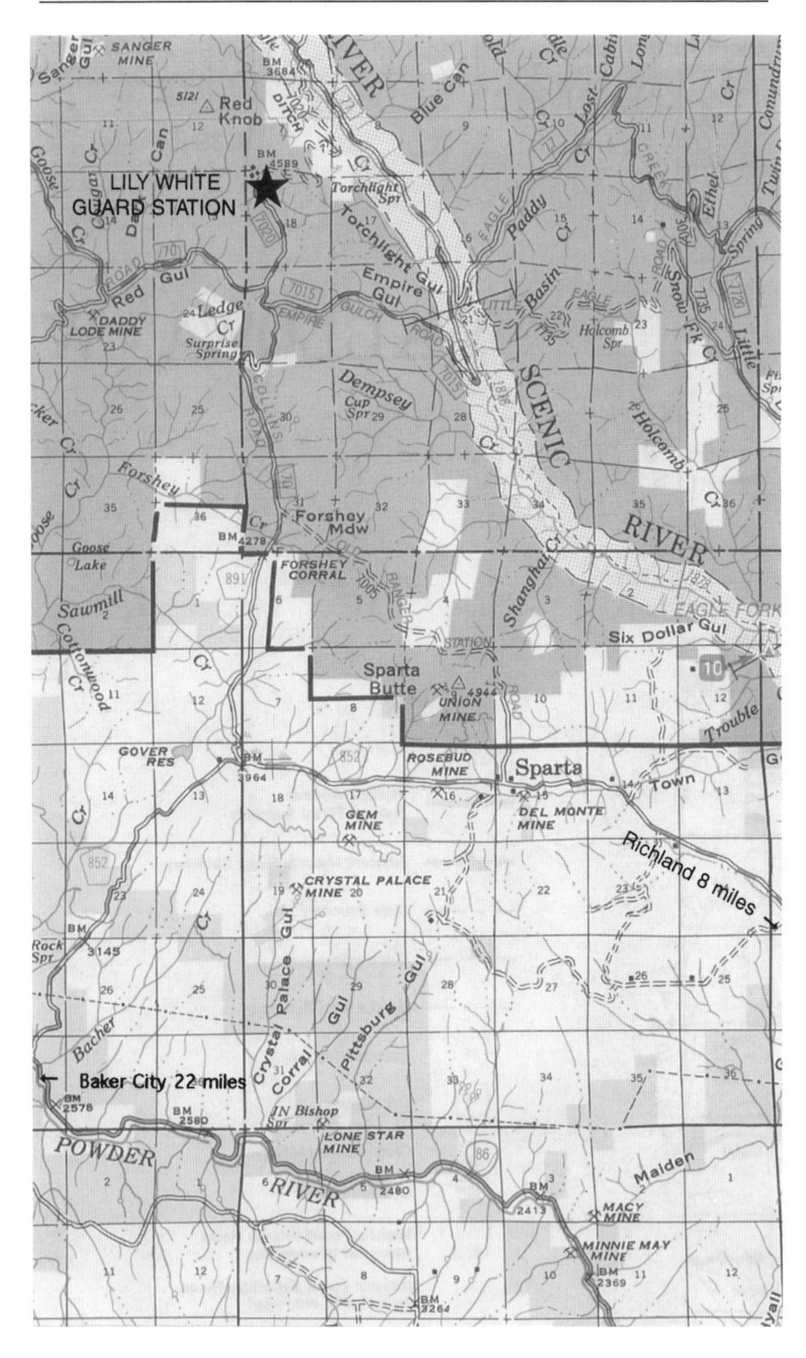

Lily White Guard Station
Wallowa-Whitman National Forest, Oregon

31 Murderer's Creek Work Center

Malheur National Forest

"I believe in the forest, and in the meadow, and in the night in which the corn grows."

— HENRY DAVID THOREAU, *"A Plea for Captain John Brown,"* 1859

Your Bearings

35 miles southwest of the town of John Day
75 miles northwest of Burns
165 miles south of Pendleton
185 miles east of Bend

Availability

November 1 through May 30. However, Bear Valley Ranger District can usually provide this or another cabin to rent any time of the year, depending on weather conditions, and they are flexible about the dates.

Capacity A maximum of four people. Ideal for families.

Description

Main cabin with kitchen and bathroom with shower. Separate sleeping quarters with two bedrooms.

Cost $25 per night.

Reservations

Available for as many as five consecutive nights. Check-in and check-out time is 2:00 PM. For an application packet, maps, and further information contact:

Bear Valley Ranger District
528 East Main Street
John Day, OR 97845
541-575-2110

How To Get There

The Izee Highway, County Road 63, is 17 miles south of the town of John Day, off Highway 395, or about 55 miles north of Burns. Take the Izee Highway west off Highway 395 for six miles to Forest Road 21. Turn right onto Forest Road 21, and continue for 12 miles, following the signs for MURDERER'S CREEK. It is paved all the way except for the final two miles. The guard station is on the left.

Forest Road 21 may or may not be plowed so, in the winter, you may require skis or snowshoes to get to the cabin. Trip conditions can be moderate to difficult depending on snow accumulation. Check with the Bear Valley Ranger District for current road and weather conditions.

Elevation 5000 feet.

Map Location Malheur National Forest, Township 15 South, Range 29 East, Sections 18 and 19.

What Is Provided

Woodstove, propane cook stove, propane light, table and chairs, four single beds with mattresses, a couch, and a chest of drawers.

There are two structures here, the cabin, and the sleeping quarters. They are rented simultaneously. The cabin, built in

1906, is thought to be the oldest Ranger Station in the Pacific Northwest—your sausages will be sauteed in history.

There is a good kitchen with fridge, sinks, closets, countertops and faucets. Do not drink the water, though, or you may be history too. There is a bathroom with shower in the cabin.

The sleeping quarters were built more than 75 year after the cabin, so they are purely functional. There are two bedrooms with large closets—large by cabin standards anyway.

What To Bring

Bring your own drinking water or have the means to treat the local supply.

History

The main cabin, constructed in 1906, still serves its original purpose as a summer home for fire-crews. As to the name "Murderer's Creek," perhaps some remorseful murderer sought forgiveness here by its green banks. Local lore has it that early settlers here were killed by Native Americans.

Around You

Murderer's Creek is murmuring in your back yard on its way west to meet the South Fork John Day River. The Aldrich Mountain/Murderer's Creek Wildlife Area, an undesignated wilderness, is nearby. Riley Creek Trailhead (216A) is to the northeast, at the end of Forest Road 2190 which is off Forest Road 21, a few miles east of the cabin.

Fields Peak Trailhead (212) is to the north at the end of Spur Road 125, which is also off Forest Road 21 a few miles north of the cabin.

Trail 212 joins Trail 216, which goes around McClellan Mountain rising to 7042 feet. From Fields Peak to Riley Creek trailheads is 10-1/2 miles.

Bighorn sheep were hunted to extinction in the nearby Aldrich Mountain/Murderer's Creek Wildlife Area. In 1978 they were reintroduced: it is thought that there are as many as sixty of these animals there today. There are also pronghorn antelope, elk, mule deer, mountain lions, and, usually, not many people.

This is wild country. If you use any of these trails, take water, a topographic map and compass. Herds of wild horses are often seen around the cabin. We waited in vain for them to come and pull us away.

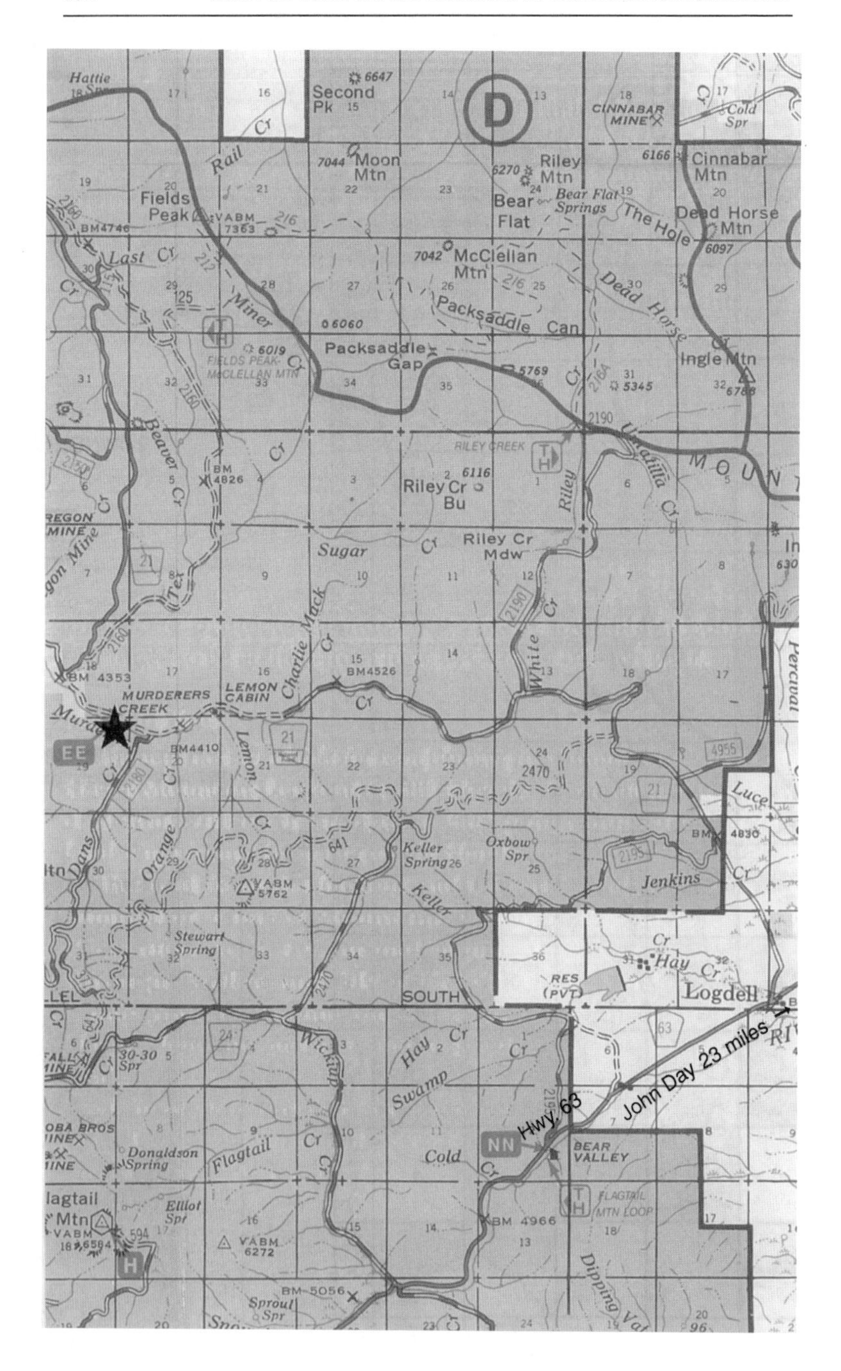

Murderer's Creek Work Center
Malheur National Forest, Oregon

32 Fall Mountain Lookout

Malheur National Forest

"To me every hour of the light and dark is a miracle,
every cubic inch of space is a miracle."

— WALT WHITMAN, *Miracles*

Your Bearings

20 miles southwest of the town of John Day
70 miles north of Burns
150 miles south of Pendleton
170 miles west of Bend

Availability Year-round.

Capacity Two people maximum.

Description 14x14-foot room atop a 20-foot tower. Electricity provided. Beautiful setting.

Cost $25 per night.

Reservations

Available for as many as five consecutive nights. Reservations are from 2:00 PM on the day you arrive until noon the day you leave. For an application packet, maps, and further information contact:

Bear Valley Ranger District
528 East Main
John Day, OR 97845
541-575-2110

How To Get There

From the town of John Day travel south for about 15 miles on Highway 395 to Starr Campground. Turn right here onto Forest Road 4920 where you will then see the sign: FALL MOUNTAIN LOOKOUT 4. Travel northwest 3-1/2 miles to the very steep and unmaintained Forest Road 607 and turn left. The lookout is one mile ahead on this rough and steep road. There is parking for a couple of cars.

There is an alternative route off Highway 395. A few miles north of Starr Campground and just south of the Vance Creek rest area, turn right onto Forest Road 3920 and continue to Forest Road 607.

In winter it may be necessary to ski or snowshoe the final 4-1/2 miles from Highway 395. Access can range from moderate to difficult, depending on the snow conditions. Consult the District Office regarding current road and snow conditions.

Elevation 6000 feet.

Map Location Malheur National Forest, Township 15 South, Range 31 East, Section 6.

What Is Provided

This lookout is the only one in this rental program with electricity—the fridge, stove, heater and lights are all electric. There are chairs and a table, double bed, closets, fire extinguisher, shovel, maps for the area, and a vault toilet.

What To Bring Your own water, and extra food.

Setting

This is a beautifully situated lookout tower, though the sense of remoteness and isolation does suffer rather badly from a microwave thingamabob that looks like the work of deranged Martians. Unfortunately, it is only about fifty yards away from the lookout. But it is an ill-wind that doesn't blow some good: without the microwave thingamabob the lookout would not have electricity.

History

Built in 1933. Back then there were miles of telephone wire stretched all the way from the lookout, through the woods, to the firefighters at Bear Valley Work Center.

Around You

From the top of the tower the cabin offers the most astonishing views of Bear Valley and the Strawberry Mountains—views that reach into tomorrow. On the ground below is a glorious old juniper tree growing right in front of the lookout. There is also a fine old wooden shed that is open to skiers in the winter season.

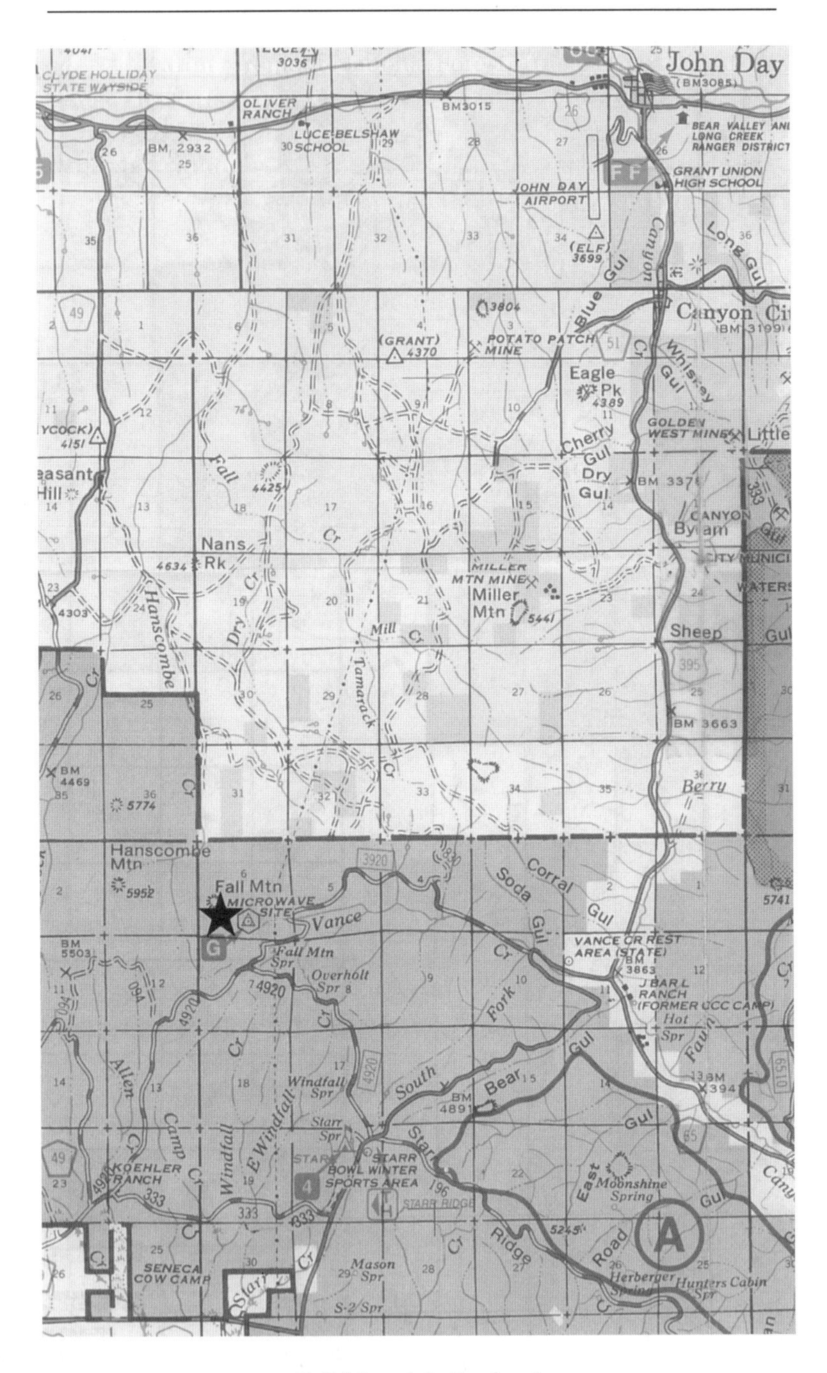

Fall Mountain Lookout
Malheur National Forest, Oregon

33 Dry Soda Lookout

Malheur National Forest

"There's no water here but plenty of opportunity for reflection."

— THOMAS DOTY, *Journal*, OCTOBER 28, 1995

Your Bearings

20 miles south of the town of John Day
65 miles north of Burns
100 miles southwest of Baker City
150 miles south of Pendleton

Availability

November 1 to June 12 are the official dates. However, Bear Valley Ranger District can provide another rental almost any time of the year, and they are flexible about the dates.

Capacity Two people maximum.

Description

14x14-foot room atop a 55-foot tower. The stairs up to the cabin are steep, especially the final flight (which, of course, is the first flight on the way down). We found it safest to go up and down facing the steps—as on a ladder.

Cost $25 per night.

Reservations

Available for as many as five consecutive nights. Reservations are from 2:00 PM on the day you arrive until noon the day you leave. For an application packet, maps, and further information contact:

Bear Valley Ranger District
528 East Main
John Day, OR 97845
541-575-2110

How To Get There

There are three routes: two off Highway 395 and one off Road 65, which is the closest paved road to the lookout.

The first route: go 16 miles south from the town of John Day, Oregon, on Highway 395 to Forest Road 196, which is on the east side of 395 at Starr Campground. It is unmarked and is easily passed, since it is close to a bend. Turn left on 196 and follow this gravel, washboarded road for eight miles, to where it joins Forest Road 3925. One mile on Forest Road 3925 brings you to Forest Road 820. Turn left. The lookout is about 100 yards ahead.

The second route off Highway 395 is another few miles past Starr Campground and is just past Highway 63. It is Forest Road 3925, off the east side of the Highway. This is also a gravel road. Go about seven miles east to Dry Soda Lookout.

The third route: go 10 miles south from the town of John Day, Oregon, on Highway 395 to Road 65—which leads southeast off Highway 395. Turn left onto Road 65 and go four miles to Forest Road 336. Turn right onto Road 336; from here it is only four gravel miles to the lookout.

Access can range from moderate to difficult, depending on the snow conditions. Consult the District Office regarding current road and snow conditions.

Elevation 5593 feet.

Map Location Malheur National Forest, Township 16 South, Range 32 East, Section 5.

What Is Provided

Propane fridge, cook stove, heater and lights. Also a double bed, table and chairs, closets, fire extinguisher, shovel, maps for the area, and a vault toilet.

What To Bring

Your own water or have the means to treat the local supply.

History

The lookout tower and cabin were built in 1941.

Around You

To the north and northeast are the snow-covered Strawberry and Canyon mountains, and Indian Springs Butte. To the south and southwest are magnificent views of Bear Valley. To the west are Flagtail and Aldrich Mountains.

There is a fine network of hiking trails a few miles north. Trailheads closest to the lookout are at the end of roads that travel north from Forest Road 65: Road 6510 leads to the boundary of Strawberry Mountain Wilderness, Road 612 leads to the trailhead for East Fork Canyon Creek Trail, and Road 1520 leads to Buckhorn Meadow and Trail 205.

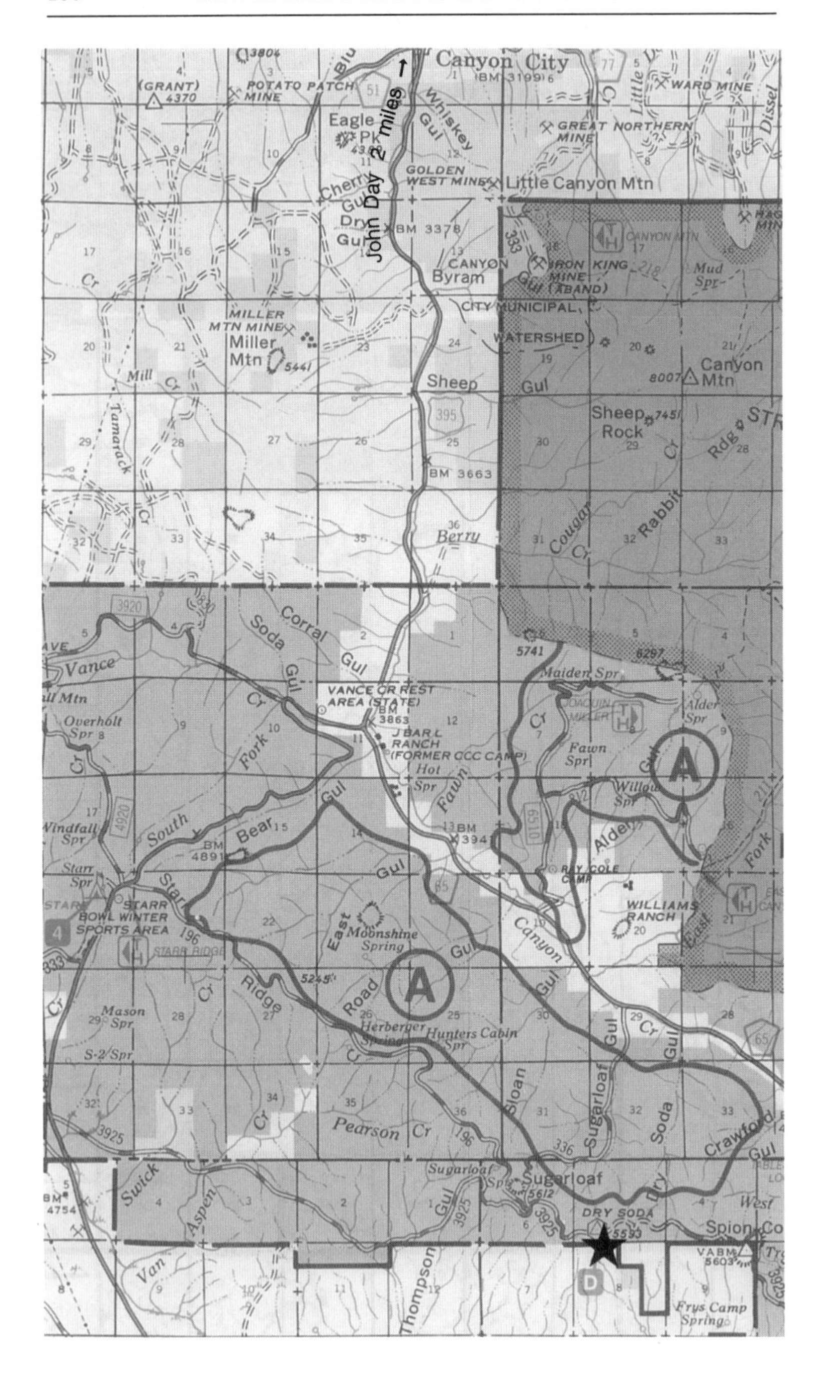

Dry Soda Lookout
Malheur National Forest, Oregon

34 Crane Prairie Guard Station

Malheur National Forest

"The strongest and sweetest songs are yet to be sung."

— WALT WHITMAN, *Boughs*

Your Bearings

30 miles southeast of Prairie City
45 miles southeast of the town of John Day
100 miles southwest of Baker City
170 miles southeast of Pendleton

Availability January 1 through April 15.

Capacity A maximum of four people.

Description A one-story, wooden cabin with two bedrooms, living room, kitchen, and breakfast nook.

Cost $25 per night, plus a refundable $25 deposit.

Reservations

Available for as many as five consecutive nights. Check-in and check-out time is 2:00 PM. For an application packet, maps, and further information contact:

Prairie City Ranger District
PO Box 337
Prairie City, OR 97869
541-820-3311

How To Get There

Travel east on Highway 26 through the center of Prairie City. Turn right at the Chevron Station onto Main Street South, then left at the T-Junction. This becomes County Road 62 and eventually joins Forest Road 16 at Summit Prairie, where you will continue southeast on Road 16.

There are no signs anywhere along the way to indicate Crane Prairie Guard Station. About 28 miles south of Prairie City on Road 16, and one mile past the junction of Forest Road 14, is Forest Road 656. It is not signed. Turn right. The Guard Station is less than a mile down Road 656.

As you go through the last of three green gates you will see a cluster of several buildings. Go past the two sets of cabins on the right as well as the cabin on the left, then take the driveway to the right. There are logs on either side of it, though, of course, everything may be covered in snow. In any event the rental cabin is the one at the end of this driveway, on the left. In front of it are two picnic tables, a barbecue and a firepit. If you look very closely you will see that the number on the cabin is 1008.

Road 62 is plowed throughout the winter, and occasionally Forest Road 16 is plowed from its junction with Road 62 to its junction with Road 14. Parking is provided at Summit Prairie and near the junction of Forest roads 16 and 14. Access conditions range from moderate to difficult. Always check with the Ranger District for the latest road reports.

Elevation 5370 feet.

Map Location Malheur National Forest, Township 16 South, Range 35 East, Section 24.

What Is Provided

Unfortunately the propane-powered fridge, stove, and lights are not available for guest use, though the Ranger District does provide wood for the woodstove. The indoor plumbing is not available for use either—it is turned off in winter to prevent pipe-freeze, but there is an outhouse near the building. There are four single beds with mattresses, as well as a couch, table and chairs, and a chest of drawers.

What To Bring

Bring your own drinking water or have the means to treat the local supply.

History

The cabin was constructed in 1932 as the ranger's house when Crane Prairie was an administrative site. There are several other buildings at this station, but this is the only one available for rent. The guard station is still in use during the summer months for seasonal fire-patrol crews.

Around You

Fringed by forested hills, Crane Prairie, probably a couple of miles long and a mile wide in places, is literally at your back door. In addition, the trailhead for Crane Creek Trail (301A) is just to the east of the cabin off Forest Road 1663. This follows Crane Creek to Crane Creek Camp (6 miles) and the North Fork Malheur Trail (381) which follows the North Fork of the Malheur River (12.2 miles total length).

A few miles north of Summit Prairie, off Road 62, is unmaintained Forest Road 101: from here the Skyline Trail (385) leads northwest into Strawberry Mountain Wilderness.

Directly opposite unmaintained Forest Road 101 is unmaintained Forest Road 994, from which the Starvation Trail (374) leads to Starvation Rock (2.1 miles).

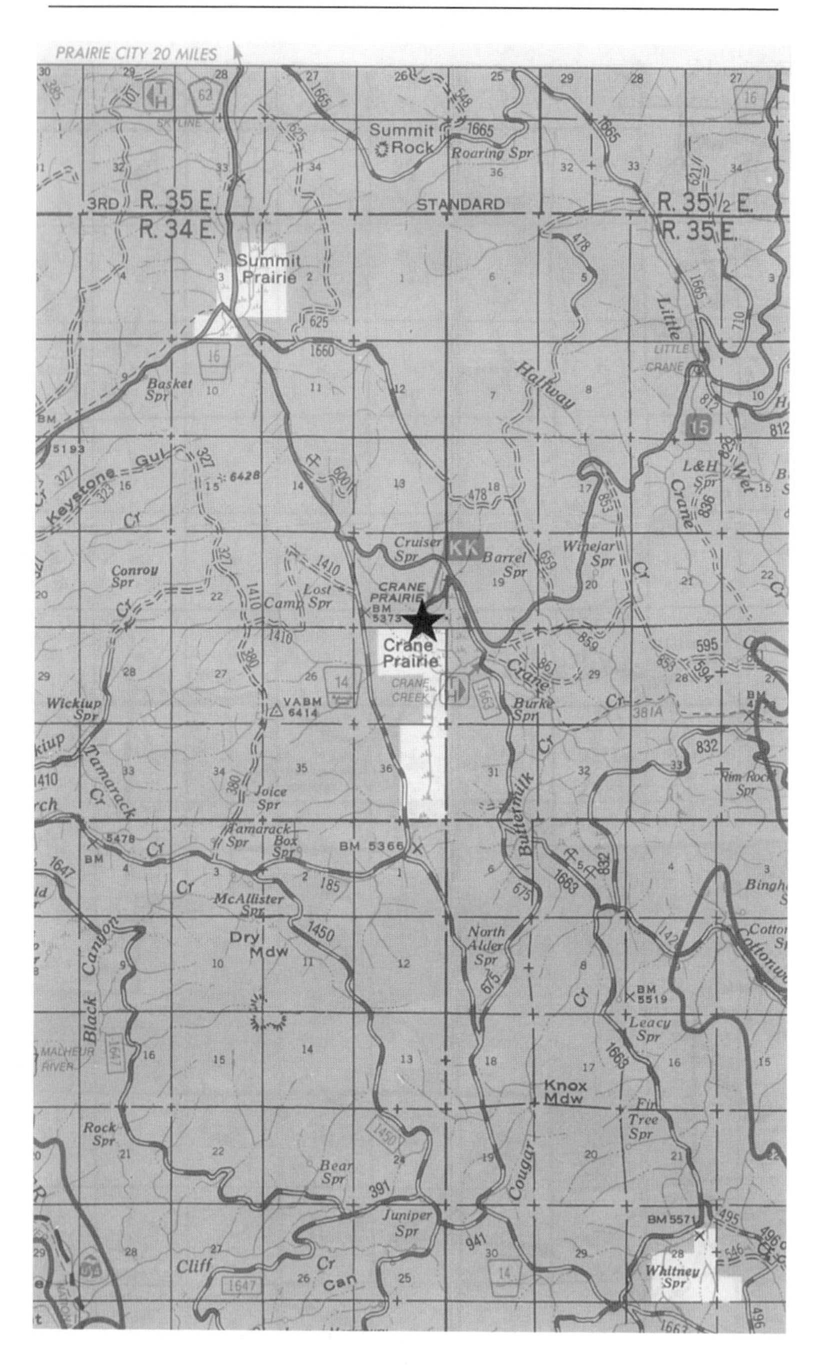

Crane Prairie Guard Station
Malheur National Forest, Oregon

35 Short Creek Guard Station

Malheur National Forest

"When we try to pick out anything by itself we find it hitched to everything else in the universe."

— John Muir

Your Bearings

25 miles southeast of Prairie City
35 miles southeast of the town of John Day
95 miles southwest of Baker City
165 miles southeast of Pendleton

Availability January 1 through April 15.

Capacity A maximum of four people. Ideal for families.

Description One-story, two-room cabin, with one bedroom and a kitchen/living room.

Cost $25 per night, plus a refundable $25 deposit.

Reservations

Available for as many as five consecutive nights. Check-in and check-out time is 2:00 PM. For an application packet, maps, and further information contact:

Prairie City Ranger District
PO Box 337
Prairie City, OR 97869
541-820-3311

How To Get There

Travel east on Highway 26 through the center of Prairie City. Turn right at the Chevron Station onto Main Street South. Turn left at the T-junction. This becomes County Road 62. Follow this paved road through rolling meadows for eight miles to Forest Road 13. Turn left and follow this road along the delightful Deardorf Creek for 18 miles. The Guard Station is on the right, at the junction of Forest Road 16.

Road access during the snow season can sometimes end at the junction of Road 62 and Forest Road 13—eighteen miles away from the guard station. Skiing conditions can range from moderate to difficult, depending on the snow and weather. Always check with the Ranger District for the current road report.

Elevation 5300 feet.

Map Location Malheur National Forest, Township 15 South, Range 35 East, Section 15.

What Is Provided

Propane heater and cook stove, four single beds, chairs, a couch, a chest of drawers, and a coffee table. Though the cabin has modern plumbing, the water is turned off and the drains winterized during the rental season to prevent pipe-freeze. Guests must use the outhouse adjacent to the cabin.

What To Bring

Drinking water or have the means to treat the local supply. Cleaning and washing water can be obtained from streams or melting snow.

History

Originally constructed in 1954 at Crane Prairie and moved to its present location at Short Creek in 1970, the cabin is still in use as a summer residence for seasonal fire-patrol crews on the Prairie City Ranger District.

Around You

Short Creek is virtually at your door, as is the North Fork of the Malheur River. A few miles to the east, via Forest Road 16 and unmaintained Forest Road 430, is Monument Rock Wilderness. This can also be reached from a few miles north on Forest Road 13 via Forest Road 1370—to the 7-1/2 mile Elk Flat Creek Trail (366).

This wilderness is home to the famously ferocious wolverine, known even to the academic world as Gulo Gulo—"glutton glutton". It eats everything that moves, including porcupines, and though only 31-44 inches long and weighing no more than 40 pounds or so, it chases off coyotes, bears and cougars from their kill. It also eats stuff that does not move, like eggs, roots and berries; or moves too slowly in the snow, such as moose or elk.

To the west are Glacier and Lookout mountains. The trailhead for Trail 371 is a mile or so north of the cabin, on Forest Road 13. It follows Sheep Creek all the way to Lookout Mountain, and, from there, you can return on Horseshoe Trail (363) to Forest Road 13, about four miles north of the cabin.

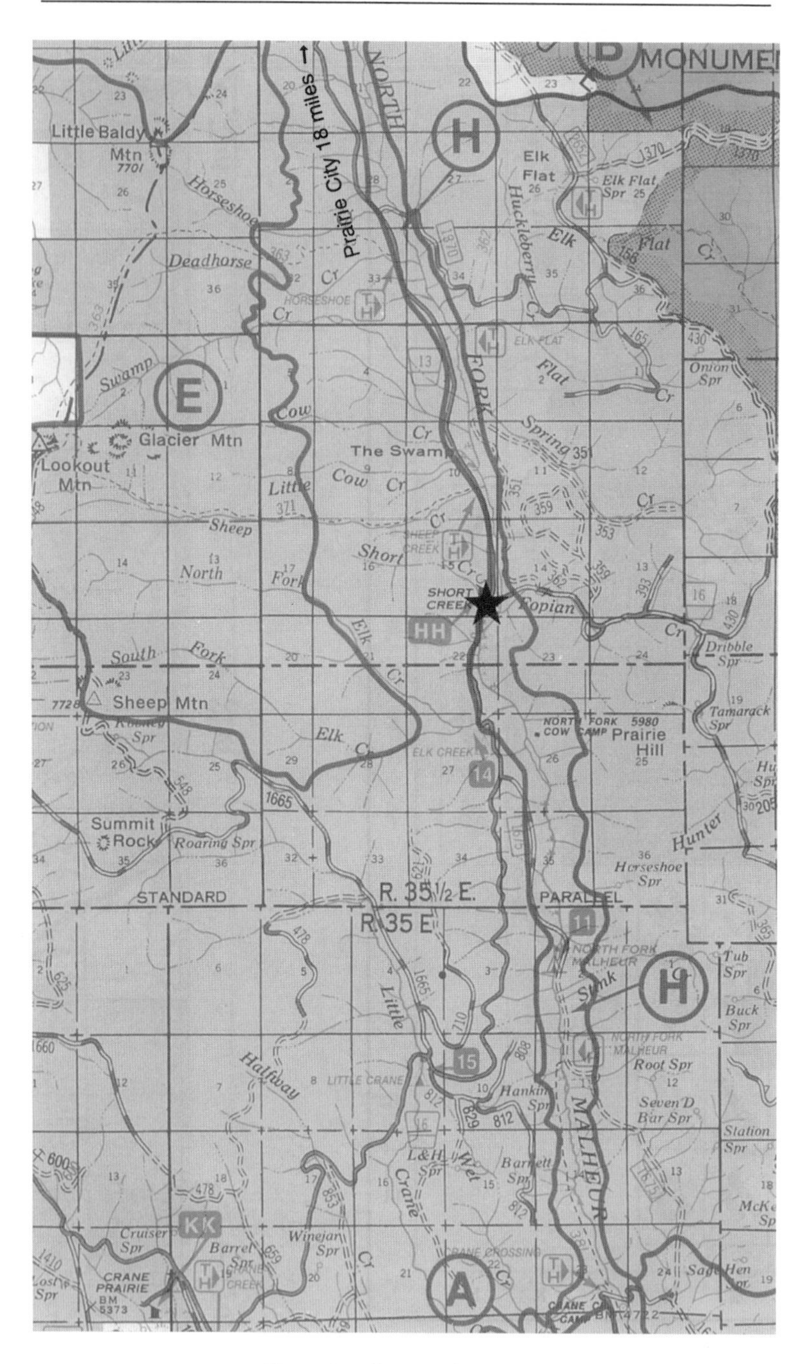

Short Creek Guard Station
Malheur National Forest, Oregon

36 ❧ Bear Valley Work Center

Malheur National Forest

"Give me the splendid silent sun with all his beams full-dazzling."

— Walt Whitman, *Miracles*

Your Bearings

25 miles southwest of the town of John Day
65 miles northwest of Burns
155 miles south of Pendleton
175 miles east of Bend

Availability

November 1 through May 31. However, Bear Valley Ranger District can usually provide this or another cabin to rent any time of the year, and they are flexible about the dates.

Capacity A maximum of four people in each of the two cabins. Ideal for families.

Description Big House: two bedrooms, living room, kitchen, bathroom. West Cabin: bedroom, kitchen, bathroom.

Cost $25 per night, plus a refundable $25 deposit.

Reservations

Available for as many as five consecutive nights. Check-in and check-out time is 2:00 PM. For an application packet, maps, and further information contact:

Bear Valley Ranger District
528 East Main Street
John Day, OR 97845
541-575-2110

How To Get There

The Izee Highway, County Road 63, is 17 miles south of the town of John Day, off Highway 395, or about 55 miles north of Burns. Take the Izee Highway west off Highway 395. Bear Valley Work Center is approximately nine miles ahead, on the left, about 100 yards away from the highway. This last 100 yards you may have to walk in the snow season. Parking is provided on Road 63. Winter access is easy, even with heavy snow conditions, but always check with the Bear Valley Ranger District for access information.

Elevation 4600 feet.

Map Location Malheur National Forest, Township 16 South, Range 29 East, Section 12.

What Is Provided

The West Cabin was being refurbished during our visit and we did not see the finished job. However, it is essentially a single room with a separate kitchen and bathroom. The kitchen has a stove, fridge, sink and cupboards. The main room has a woodstove—the Forest Service supplies the wood—and bunkbeds for four, though two people would be more comfortable in a cabin of this size.

The Big House represents better value to potential guests. It has two bedrooms, each with bunk beds for two, a living room and a kitchen. The kitchen has a marvelous old six-burner propane stove, two fridges, two sinks and a woodstove. The Forest Service supplies the wood.

What To Bring

The water is turned off in the winter to prevent pipe-freeze, so you need to bring your own or have the means to treat the local supply. Check with the Ranger District before your departure.

Setting

Driving up to the parking area you will see a cluster of seven or eight buildings, all historic structures built in the 1930s. Only the West Cabin, which is on the right of the driveway and faces the parking area, and the Big House, which is the biggest of the structures and is on the extreme left, are available for rent.

History

The two cabins were constructed in 1934 as part of the original headquarters of the Bear Valley Ranger District, and they still serve the District as summer homes for fire-suppression crews.

Around You

The Silvies River flows placidly in front of the cabins and continues east through lovely Bear Valley. To the west, on Road 63, is the South Fork of the John Day River.

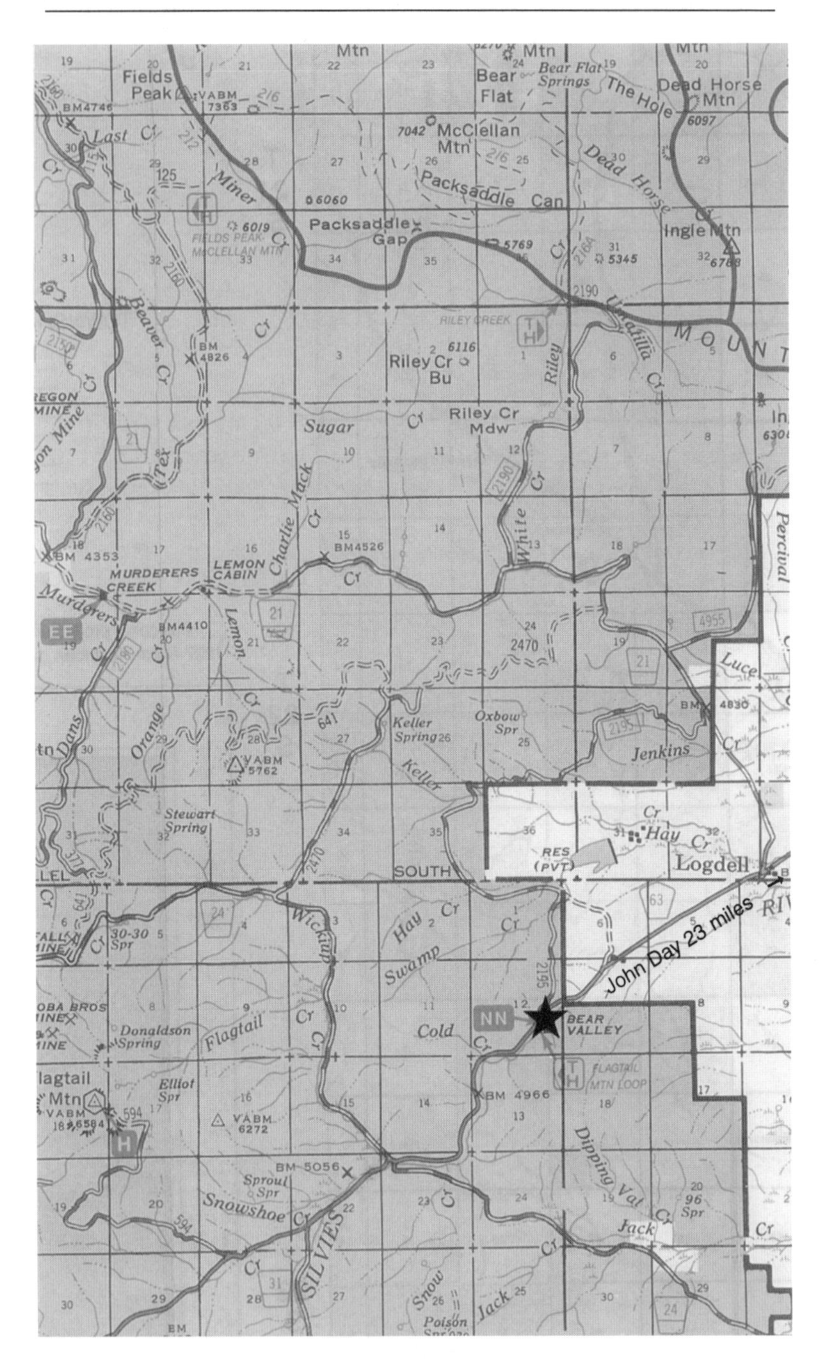

Bear Valley Work Center
Malheur National Forest, Oregon

37 Flag Tail Lookout

Malheur National Forest

"The sky is the daily bread of the eyes."

— Ralph Waldo Emerson, *Journal*, May 25, 1843

Your Bearings

35 miles southwest of the town of John Day
70 miles northwest of Burns
165 miles south of Pendleton
185 miles east of Bend

Availability

November 1 to June 6 are the official dates. However, Bear Valley Ranger District can usually provide another rental any time of the year and they are flexible about the dates.

Capacity Two people maximum.

Description

14x14-foot room perched on a 60-foot tower, with catwalk, 73 steps up from the ground. Designed for only one or two people.

Cost $25 per night.

Reservations

Available for as many as five consecutive nights. Reservations are from 2:00 PM. Check-out time is at noon. For an application packet, maps, and further information contact:

Bear Valley Ranger District
528 East Main
John Day, OR 97845
541-575-2110

How To Get There

Travel south from the town of John Day on Highway 395 for 17 miles to the Izee Highway (Road 63), or, if traveling from Burns, travel north on Highway 395 for 55 miles miles to this same junction. Take the Izee Highway west for 12 miles to signed Forest Road 594. Turn right onto Road 594. The lookout is five miles ahead, at the end of this road.

Winter access can range from moderate to difficult, depending on the snow conditions. Check with the Ranger District for a current road report prior to your departure.

Elevation 6400 feet.

Map Location Malheur National Forest, Township 16 South, Range 29 East, Section 18.

What Is Provided

Propane fridge and fuel, cook stove, heater and lights. Also a double bed, table and chairs, closets, but no sink or woodstove. The cabin is equipped with a fire extinguisher, shovel, and maps for the area. The garage downstairs is open in winter to cross-country skiers. There is a vault toilet nearby.

What To Bring

Your own water or have the means to treat snowmelt.

History

Established as a lookout site in the 1930s. The tower was rebuilt in the early 1960s, and still serves as a fire lookout during the summer for the Bear Valley Ranger District and adjoining lands.

Around You

To the west: the snow-capped Three Sisters and Mt. Bachelor; to the north: Fields Peak in the Aldrich Mountains; to the east: Strawberry Mountain, Canyon Mountain, Elkhorn Mountain and Bear Valley; to the southwest: Snow Mountain.

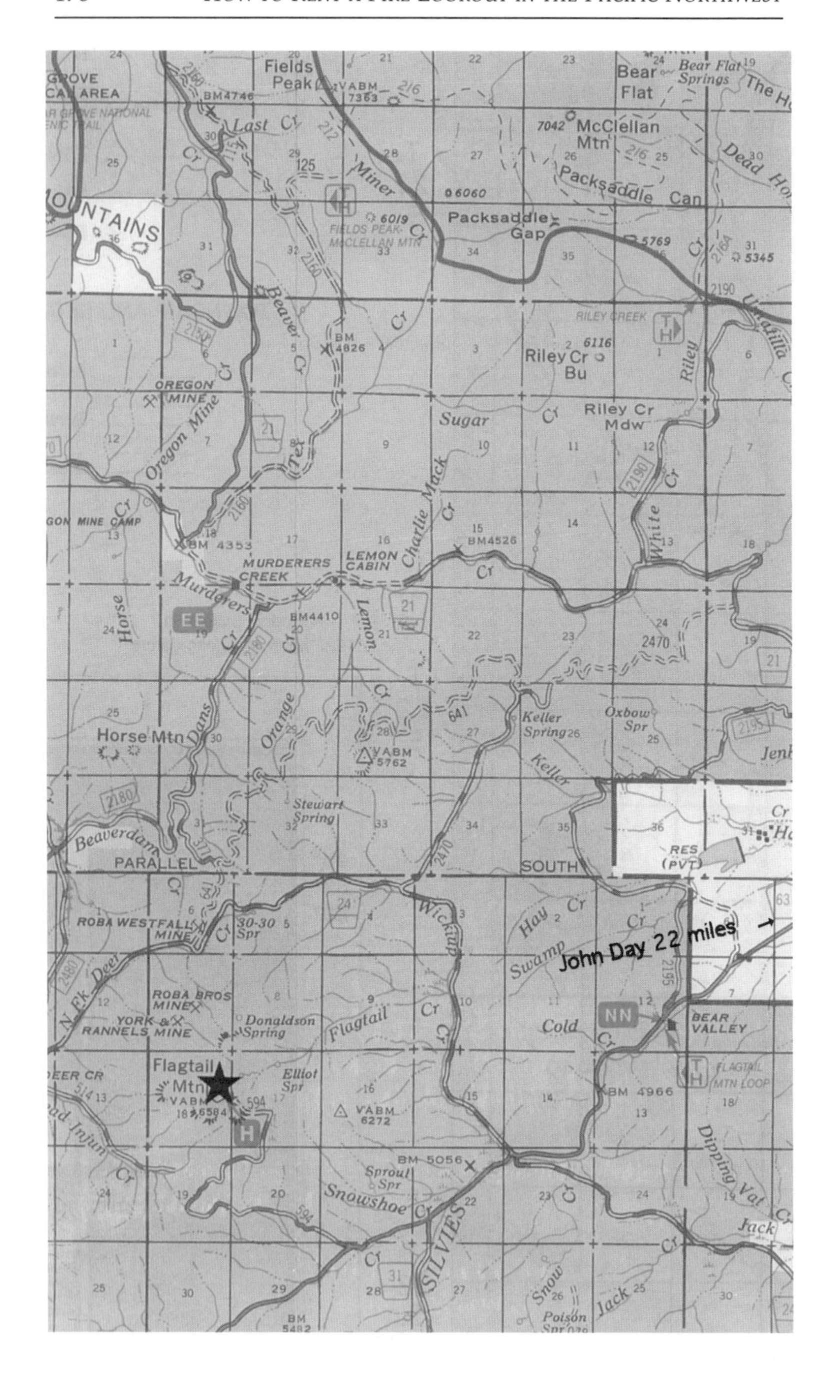

Flag Tail Lookout
Malheur National Forest, Oregon

38 ✤ Deer Creek Guard Station

Malheur National Forest

"I know of no more encouraging fact than the unquestionable ability of man to elevate his life by conscious endeavor."

— Henry David Thoreau, *Walden*

Your Bearings

35 miles southwest of the town of John Day
75 miles northwest of Burns
165 miles south of Pendleton
185 miles east of Bend

Availability

November 1 through June 6. However, Bear Valley Ranger District can usually provide this or another cabin to rent any time of the year, depending on weather conditions and they are flexible about the dates.

Capacity A maximum of four people. Ideal for families.

Description 20x14-foot, one-room cabin, with bathroom and shower.

Cost $25 per night.

Reservations

Available for as many as five consecutive nights. Check-in and check-out time is 2:00 PM. For an application packet, maps, and further information contact:

Bear Valley Ranger District
528 East Main Street
John Day, OR 97845
541-575-2110

How To Get There

The Izee Highway, County Road 63, is 17 miles south of the town of John Day off Highway 395, or about 55 miles north of Burns, off Highway 395. From here take the Izee Highway west for ten miles to Forest Road 24. Turn right, and follow Forest Road 24 north and then southwest for another nine miles to Forest Road 514. Turn left. The guard station is on the right, less than a half mile down this road. You will find plenty of parking space.

In winter, Forest Road 24 may not be plowed, so skis and snowshoes may be necessary. Access under such conditions can be moderate to difficult depending on snow accumulation. Check with the Bear Valley Ranger District for current road and weather conditions.

Elevation 5100 feet.

Map Location Malheur National Forest, Township 16 South, Range 28 East, Section 14.

What Is Provided

The inside of the cabin has lovely tongue-and-groove pine paneling, high ceilings, curtained windows, vinyl floor, fridge, stove, sinks, closets, tables and chairs, chests of drawers, two single beds and two couches.

Also provided are propane lights, a propane heater and a flush toilet and shower. There is a vault toilet outside, and a shed.

What To Bring

In winter the water is turned off to prevent pipe-freeze, so bring your own drinking water, or have the means to treat the local supply.

Setting

This small cabin is beautifully situated in a huge meadow that contains a massive ponderosa pine, said to be one of the biggest in the Malheur National Forest. Deer Creek is at one end of the meadow, on its way west to join the South Fork John Day River, and Dead Injun Creek babbles by the firering and barbecue area, on its way to meet Deer Creek.

History

In 1956 the Forest Service built the guard station here to house a fire guard.

Around You

We noticed a surprising amount of old-growth trees between the cabin and Flag Tail Lookout, which is a couple of miles east. This entire area is a maze of old logging roads. You will surely find a gold mine down one of them—some who came here before you did.

The cabin is on the brink of Aldrich Mountain/Murderer's Creek Wildlife Area, where you may encounter bighorn sheep, pronghorn antelope, mule deer, elk and mountain lions.

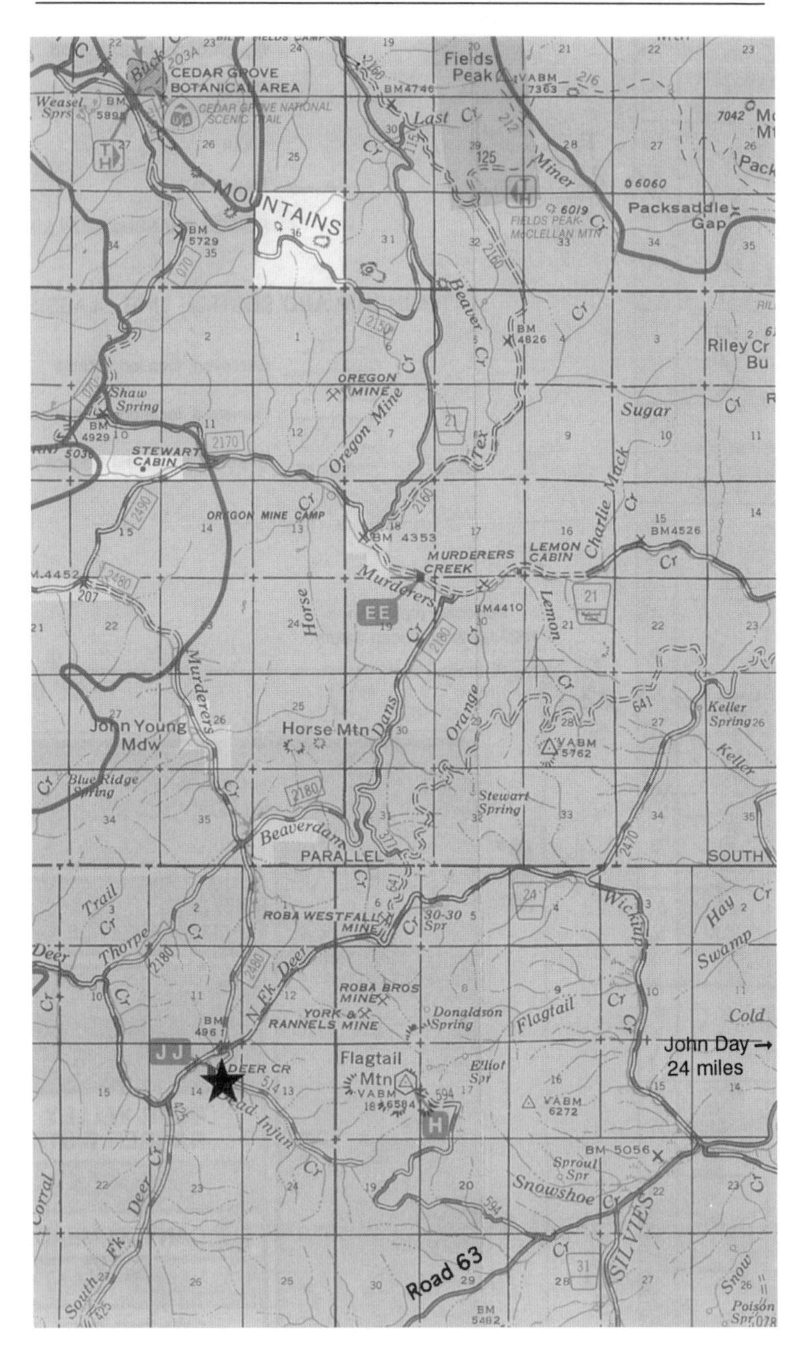

Deer Creek Guard Station
Malheur National Forest, Oregon

39 Hager Mountain Lookout

Fremont National Forest

"The ultimate finale was the parting of the clouds and a vivid red sunset over Mt. Shasta."

— From the Lookout's Guestbook, 1994

Your Bearings

15 miles south of the town of Silver Lake
60 miles north of Bly
95 miles south of Bend
100 miles northeast of Klamath Falls

Availability November 15 through May 15.

Capacity Maximum of four, though two would be more comfortable.

Description 14x14-foot room with a jaunty hip roof and deck. In need of some T.L.C., but with incomparable views.

Cost $25 per night.

Reservations

Available for as many as seven consecutive nights. For an application packet, maps, and further information contact:

Silver Lake Ranger District
PO Box 129
Silver Lake, OR 97638
541-576-2107

How To Get There

From the Silver Lake Ranger District office, travel 1/2 mile east on Highway 31, then turn right (south) onto County Road 4-12, which becomes Forest Road 28. Take Road 28 for 11 miles, then turn left onto Forest Road 036. This gravel road is not signed until you have turned onto it, and, since it is at a bend in Forest Road 28, it is very easily passed by. As a guide, it is two miles south of the TRAIL sign and about 100 yards south of the SHARP CURVES 25 MPH sign. Turn left onto Road 036 and continue to Forest Road 497, about two miles ahead.

Here you will notice a sign: HAGER MOUNTAIN LOOKOUT 3—but don't believe it, it is over four miles. Despite signs to the contrary, many cars, at least in summer, should be able to negotiate this road for another 2-1/2 miles to the green gate. Here, though, all cars should park. The final mile or so to the lookout from the locked green gate is extremely steep with very sharp hair-pin bends, and should not be attempted in a vehicle that does not have four-wheel drive. Visitors on foot, snowshoes or skis should be prepared for an arduous climb.

From Bly travel three miles west on Highway 140 to Ivory Pine Road, also County Road 1257, which becomes Forest Road 036. To continue on pavement, follow this road all the way to Forest Road 28—about 30 miles ahead. Turn left onto Forest Road 28 and continue north for about 24 miles, following the Silver Lake signs, to Forest Road 036. This gravel road is not signed; it is on the right 3-1/2 miles north of the sign: EAST BAY CAMPGROUND. Turn right onto 036 and continue for one mile to Forest Road 497, which is the access route (above) to the lookout.

For hikers there are a few alternatives: one is from a trailhead which is nine miles south of Highway 31 on Forest Road 28. It is signed and has parking. The lookout is four miles away—but be forewarned, this is a strenuous hike by most people's standards and may be too strenuous for some.

Another option is to turn left on unmaintained Forest Road 012, which is off of Road 28 but is not signed—it is close to the trailhead mentioned above. Continue on this road for about two miles, until it ends. The lookout is now only two miles away, but, again, take heed, that two miles has an elevation gain of 1600 feet. That may be about 1000 feet too much for some, but for others a welcome challenge.

Before approaching the lookout in winter, it is essential to check with the Ranger District to determine your best method of travel. Snow levels are apt to vary greatly from one year to the next, even one day to the next. Skiing or snowshoeing from the parking area can take as long as half a day depending on weather conditions and one's ability.

Elevation 7200 feet.

Map Location Fremont National Forest, Township 29 South, Range 14 East, Section 35.

What Is Provided

Two cots, two beds, two fridges, a very small woodstove and firewood, also a propane cook stove, two large pots for boiling water and a coffee pot, but no sink. An outhouse, horse tie-racks and picnic table are close to the cabin but, according to some of the comments in the guestbook, the outhouse is not nearly close enough.

Renting this lookout in winter is definitely for the adventurous, and even those who consider themselves adventurous should also be fit and enjoy coping with snow drifts, gale-force winds, and the possibility of nocturnal visits to an outhouse that is about 100 yards away.

What To Bring

Drinking water is a must—none is provided. Snowmelt can be used for your washing needs, though safe drinking water from snow cannot be assured, so have the means to treat it. Bring extra food—severe weather conditions may delay your departure. And bring those binoculars, even if they do add a pound or two. You will have a fair amount of land and sky to survey.

Setting

Hager Mountain Lookout is one of a diminishing number of mountaintop sites still staffed by fire-guards throughout the summer and early fall. Although a trip to the summit is a very rewarding experience at any time of the year, the lookout may be rented only when the threat of wildfire has passed and the seasonal lookout personnel have returned home. So, dust off your skis; there are open slopes below the lookout, which in good snow years provide some spectacular telemark skiing opportunities.

History

The original Hager Mountain Lookout was an L-4 ground-mounted type constructed in the 1920s from a kit. Its replacement was built in the 1960s. Better road access, reconnaissance flights, and satellite detection systems have decreased the number of fire lookouts in use on National Forest and State protected lands, but Hager Mountain Lookout continues to be staffed from June to October of each year.

Around You

Though not available for rent during the summer or early fall, Hager Mountain Lookout is still a lovely destination. On a clear day you will be treated to a 360° panoramic view of the landscape from Mt. Shasta to Mt. Hood.

The trail from the lookout ends about 26 miles west at the summit of Yamsey Mountain, which is a Semi-Primitive Recreation Area, i.e., an undesignated wilderness, passing through Antler and Silver Creek Marsh Trailheads and Campgrounds. Thompson Reservoir is within five miles of Hager Mountain and has camping facilities on its east and west banks.

Near the lookout is a gravestone surrounded with flowers, placed here as a memorial to a former Hager Mountain lookout guard who died in a car crash.

"Goodbye to the happiest days I have spent on this earth. Thank you Hager Mountain. I will never ever forget the great times, the loving people and the breath-taking views."

— FROM THE LOOKOUT'S GUESTBOOK

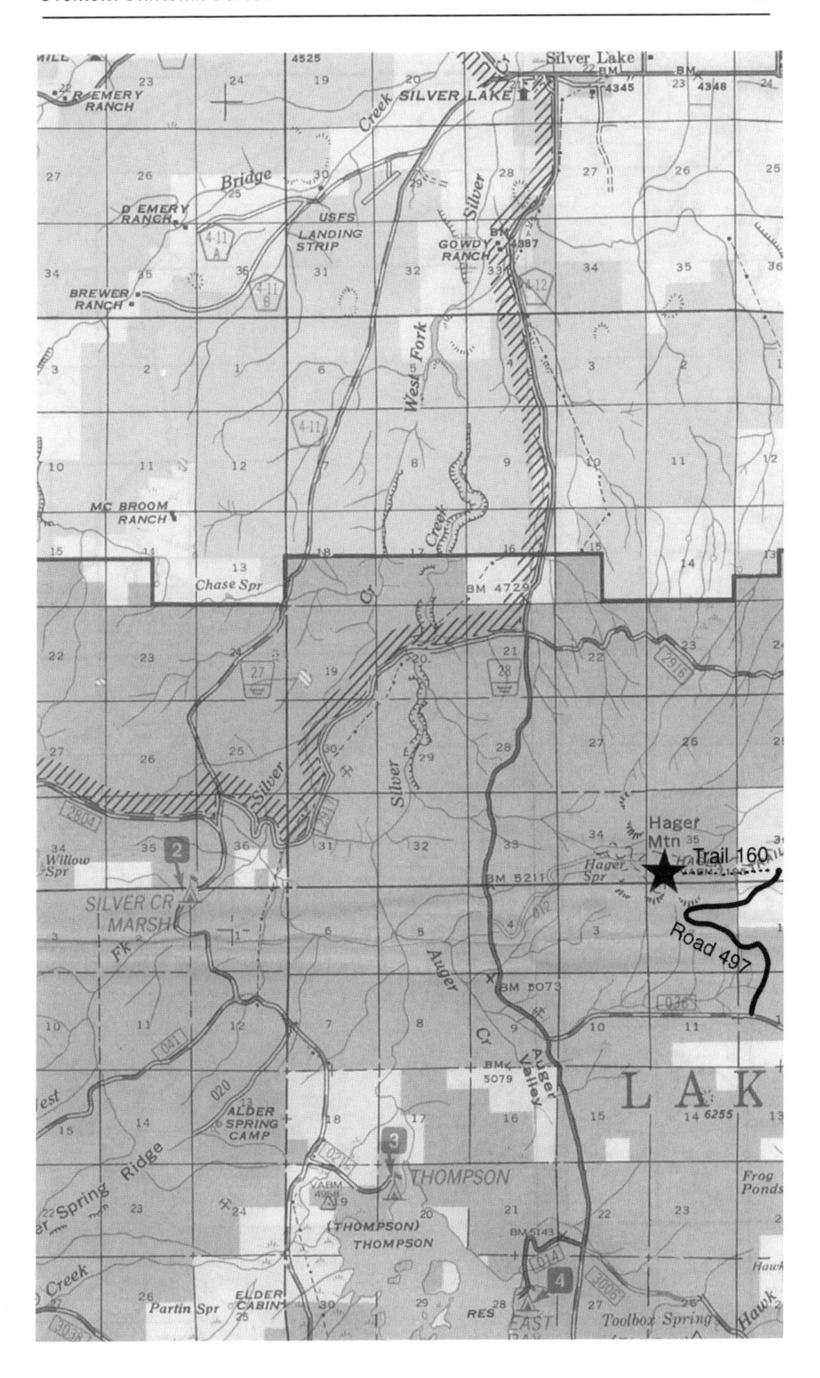

Hager Mountain Lookout
Fremont National Forest, Oregon

40 Bald Butte Lookout

Fremont National Forest

"I have a great deal of company in my own house; especially in the morning when nobody calls."

— Henry David Thoreau, *Walden*

Your Bearings

25 miles west of Paisley
40 miles northeast of Bly
85 miles northeast of Klamath Falls
110 miles south of Bend

Availability Year-round.

Capacity Four people, though two would be more comfortable. The cabin and outhouse are wheelchair-accessible.

Description 14x14-foot room with hand-made furniture and extraordinary views.

Cost $25 per night.

Reservations

Available for as many as five consecutive nights. Check-in and check-out time is 2:00 PM. For an application packet, maps, and further information contact:

Paisley Ranger District
Paisley, OR 97636
541-943-3114

How To Get There

The final 1-1/2 miles of dirt road to Bald Butte Lookout are navigable by low-clearance highway vehicles—but only barely—even in summer. The road is steep, rough and rocky in places. Leave the Ferrari at home, and don't get distracted by those lovely groves of aspen that fringe the rolling meadows along the way.

From Paisley Ranger Station, travel 1/2 mile north on Highway 31, and turn left onto Mill Street, which is County Road 2-8. Follow Mill Street to a "Y" intersection and veer right onto Forest Road 3315. Continue for 18 miles to Forest Road 28. Turn left. Then, travel less than a mile to the junction with Forest Road 3411.

Turn right onto Forest Road 3411, following the sign: LEE THOMAS CAMPGROUND, SANDHILL CAMPGROUND, BLY, and travel 1–1/2 miles on gravel to Forest Road 450—which is just before the Whitehorse Creek crossing. Turn right up the dirt road. Bald Butte Lookout is 1-1/2 miles ahead.

From Bly Ranger Station (not shown on our map), travel east on Highway 140 for one mile. Turn left (north) at the sign for Campbell Lake and proceed for about one half mile to Forest Road 34. Turn right and drive 19 miles to Forest Road 3372. Turn left and travel 10 miles to Forest Road 3411. Here, turn right. Travel four miles to Forest Road 450, which will be on your left just after you cross Whitehorse Creek. Take Forest Road 450 for 1-1/2 miles up to the lookout.

Winter visits, from November to May, will likely require over-the-snow access for the final 1-1/2 miles via skis, or snowshoes. But the view is definitely worth the trip.

Elevation 7536 feet.

Map Location Fremont National Forest, Township 34 South, Range 16 East, Section 14.

What Is Provided

The lookout itself is immaculately clean and has been lovingly restored to its 1930s days of glory. The furniture, including a double bed, table, four chairs, and a bench, was built from original 1920s plans. These furnishings are a delight to behold and a credit to the many people involved.

Propane cooking and heating appliances are provided, as well as a propane light. There is no fridge. An outhouse is next to the cabin.

What To Bring

Drinking water is a must—none is provided. Snowmelt can be used for your washing needs, though safe drinking water from snow cannot be assured, so have the means to treat it. Bring extra food—severe weather conditions may delay your departure.

History

Bald Butte Lookout was constructed from a kit as an L-4 Ground-Mounted Lookout. This type of lookout was once common throughout Oregon, but Bald Butte is now one of the few in the state. The lookout was assembled for a total cost of $668. The kit and materials cost $558 and the construction costs were $110. It was assembled here in a matter of days during the summer of 1931. For the next fifty years, its crew watched over Gearhart Mountain, Sycan Marsh, Yamsay Mountain, Lee Thomas Meadow, Slide Mountain, the upper Chewaucan River drainage, and all the forested country in between.

As testimony to the soundness of its original design and construction, it is still standing after sixty-plus years of almost constant buffeting by strong winds, as well as seasonal snow and rainstorms.

A Kellogg hand-crank telephone was the primary means of communication with other fire lookouts and guard stations in Fremont National Forest. Water was hauled from Bald Butte Spring, about one-half mile southwest.

During World War II, the lookout staff kept careful vigilance for Japanese fire-bombs, which were launched from Japan on hot air balloons and carried on the winds across southern Oregon. It is estimated that 9000 of these bombs were launched between November 1944 and April 1945 to spread panic by igniting massive forest fires, and thus divert America's resources from the war effort.

About 30 miles south of here, on Forest Road 34, there is a monument, the Mitchell Monument, to Elsie Mitchell and five children, who were the only Americans killed in the U.S. by enemy action during World War II. They were the unwitting victims of a Japanese balloon bomb. It happened on May 5, 1945, when Reverend Archie Mitchell took his pregnant wife, Elsie, and five Sunday school students to the woods for a picnic just a few miles northwest of Bly. While Reverend Mitchell parked the car, his wife and the children noticed an unusual object on the ground among the trees. When they went to inspect it, it exploded, killing them all. Reverend Mitchell was the sole survivor.

Records show that the lookout was maintained periodically until the mid-1960s. More sophisticated fire detection, including aerial surveillance, electronic observation, and better road access, made many lookouts obsolete. In time, most were abandoned or, worse, destroyed. Bald Butte Lookout was rescued from a similar fate in September 1993, through the efforts of two groups of volunteers in the *Passport in Time* program. They, with Forest Service staff, restored the lookout to its original 1930s condition. It is now available to you, a happy outcome for this little cabin in the clouds.

Around You

Overlooking the Gearhart Wilderness in south-central Oregon. On a clear day you can see part of California, and, on clearer days, part of Washington State—or so we are told.

When we were there in late August, the very last remnant of winter snow was still on the ground. The air in the glorious meadow around the lookout was heavy with the mingled scents of sage, pine and alpine wildflowers, while the long grass undulating in the breeze made the hillside look like an ocean.

To the west and northwest are Mt. Scott, Mt. McLoughlin and Diamond Peak. To the south and southwest are glorious Mt. Shasta, the Gearhart Wilderness and Dead Horse Rim. To the east are Warner Peak, Abert Rim, Hart Mountain, Drake Peak, and Brattain, Morgan, and Avery Buttes. To the north is Slide Mountain and on a clear day, a view into forever.

If you can bear to leave this place, the Lee Thomas Trailhead is on Forest Road 3411, a couple of miles west of the Forest Road 450 junction. From the trailhead the Dead Horse Trail goes south past Lee Thomas Meadow to Dead Horse Rim. It also leads to Dead Horse Lake and Campbell Lake.

Remnants of an old tree lookout on Flag Tail Peak.

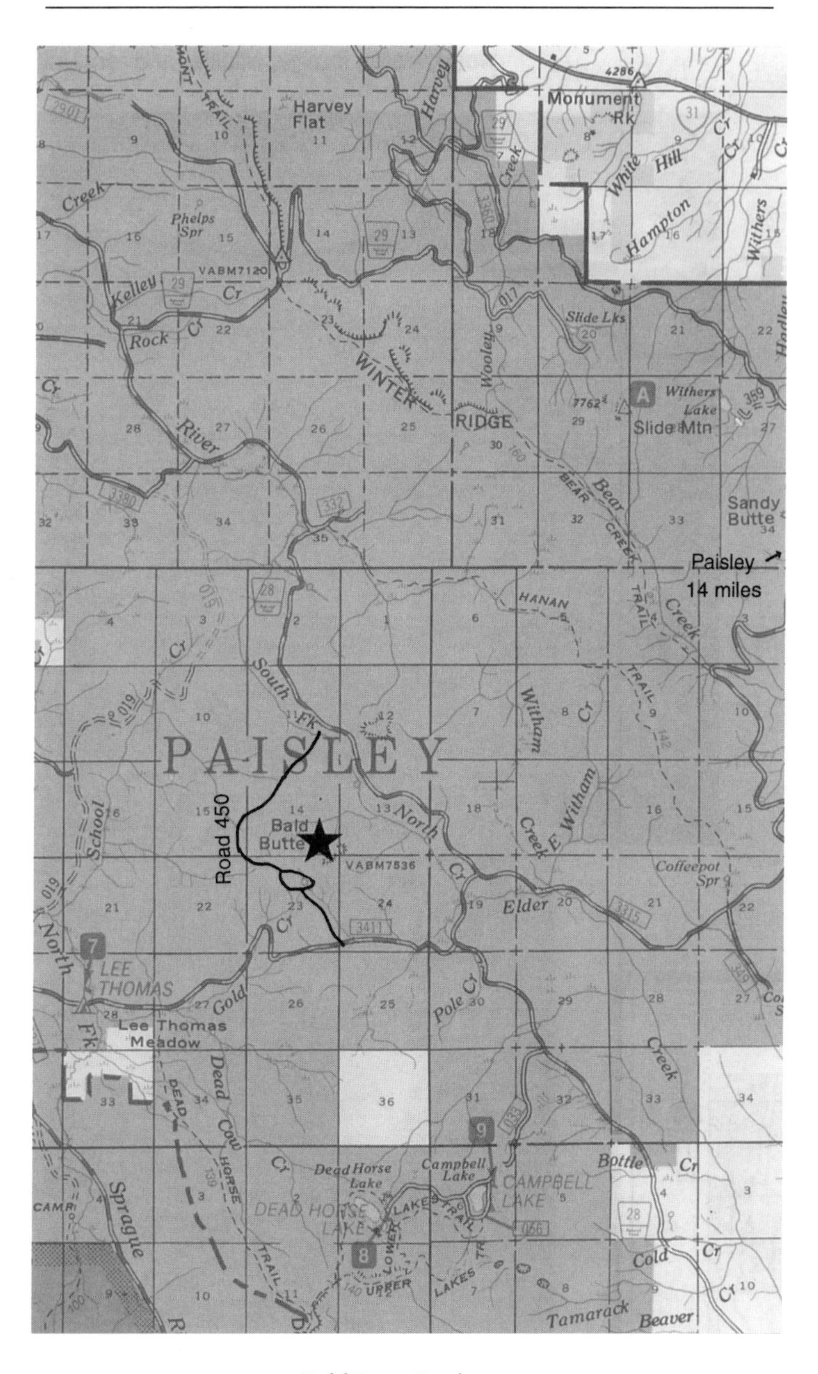

Bald Butte Lookout
Fremont National Forest, Oregon

41 Aspen Cabin
Fremont National Forest

"I believe a leaf of grass is no less than the journey-work of the stars."
— Walt Whitman, *"Song of Myself"*

Your Bearings
25 miles northeast of Lakeview
120 miles east of Klamath Falls
200 miles southeast of Bend

Availability June 1 to October 15 as an overnight rental; November through April as a free warming shelter for day use only.

Capacity Up to six people, but three would be more comfortable. Ideal for families.

Description Lovely, historic log cabin in a delightful setting.

Cost $25 per night.

Reservations

For an application packet, maps, and further information contact:

Lakeview Ranger District
HC 64 Box 60
Lakeview, OR 97630
541-947-6359

How To Get There

The route is paved all the way. From Lakeview travel five miles north on Highway 395 and turn right onto Highway 140. Go east on Highway 140 eight miles to North Warner Road (Forest Road 3615). Turn left and go north another eight miles, past aspen-fringed Bull Prairie and Mud Creek Campground. The cabin is on the right.

Elevation 6500 feet.

Map Location Fremont National Forest, Township 38 South, Range 21 East, Section 12.

What Is Provided

Drinking water, woodstove, table and chair, and a vault toilet.

What To Bring Firewood.

Setting

A log cabin enticingly situated among groves of aspen and stands of ponderosa pine and white fir. There is a heart-warming, foot-freezing little stream flowing just in front of the gate as deep as it is wide—about 12 inches.

In early July, when we were there, wildflowers and butterflies, almost indistinguishable from one another, were dancing on the banks of this stream and, indeed, all over the entire grounds. Unfortunately, in the evening, mosquitoes joined in and we too started dancing—-all the way to our backpacks for repellent.

History

This historic log cabin was built in 1930 and served as a Forest Service guard station through the late 1970s.

Around You

Abert Rim, to the north, is 2000 feet high and more than 30 miles long, which makes it the highest exposed escarpment in North America. Thermal winds off Lake Abert make possible 20-mile hang-gliding flights along the length of the rim, and make Tague's Butte a well-known launching site.

To reach Tague's Butte continue north on Forest Road 3615 another 13 miles to the *second* entrance to the looped Forest Road 032. After 3/4 mile, take the spur road through the gate for about 1/2 mile (not shown on our map). Tague's Butte juts out from Abert Rim and provides a expansive view of the valley, and Lake Abert.

On the 4th of July it is the site of the Annual Hang-Gliding Fly-In. On that weekend, in nearby Lakeview, there is not a vacant hotel room in town. Yet, on that same weekend, while people roamed the town searching in vain for a place to sleep, this delightful cabin, closer than any other accommodation to Tague's Butte, the actual hang-gliding launch site, was vacant.

To find the best view closest to the cabin, travel one mile west of the cabin and turn left onto Forest Road 019. Follow this to Drake Peak Lookout which, in fact, is on Light Peak. You can reach Drake Peak itself via Spur Road 138 (not shown on map) and hike one mile to the top.

In the vicinity of the cabin are numerous trails (not shown on our map; though trail information is available at the Ranger District). Some lead into the Drake-McDowell Semi-Primitive Non-Motorized Recreation Area. They are open to hikers, equestrians, and mountain-bikers. All have spacious trailheads for parking and trailer turnaround. All except the Swale trailhead have toilet facilities. The South Fork Crooked Creek Trail also has a horse feeder and tie racks. The Walker trail leads to the Crane Mountain National Recreation Trail.

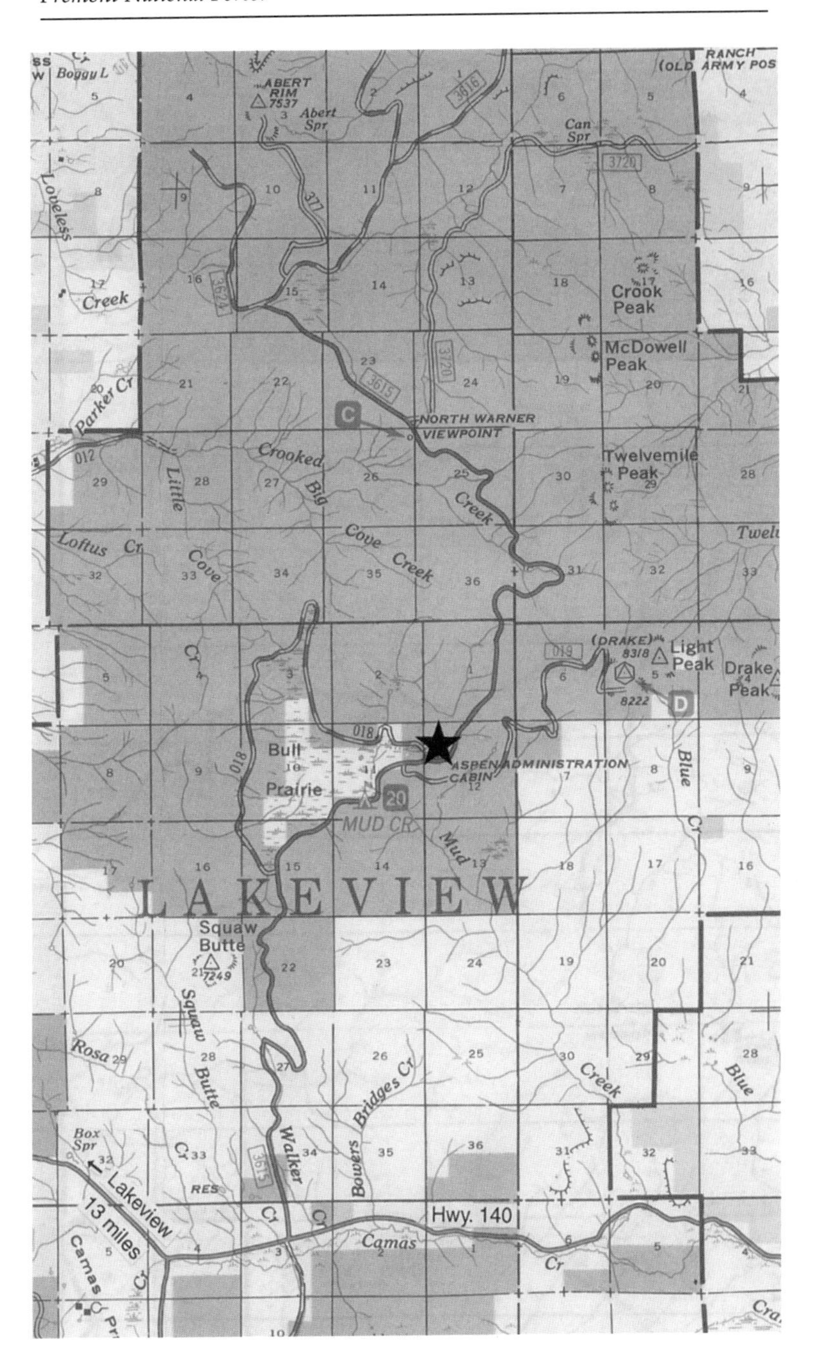

Aspen Cabin
Fremont National Forest, Oregon

42 Fremont Point Cabin

Fremont National Forest

"It is absolutely beautiful up here on the ridge. I was born and raised in Oregon and I never knew that this peaceful and serene place existed . . ."

— FROM THE CABIN'S GUESTBOOK

Your Bearings

30 miles northwest of Paisley
40 miles southeast of Silver Lake
70 miles northwest of Lakeview
115 miles southeast of Bend

Availability Year-round.

Capacity Up to four people, though two would be more comfortable.

Description 12x15-foot, one-room cabin with deck. Glorious views.

Cost $25 per night.

Reservations

Check-in and check-out time is 2:00 PM. For an application packet, maps, and further information contact:

Silver Lake Ranger District
PO Box 129
Silver Lake, OR 97638
541-576-2107

How To Get There

From the town of Paisley travel 12 miles north on Highway 31 to Forest Road 29. This is not signed and is easy to miss. Turn left here and follow this winding, steep, well-gravelled road for ten miles to Forest Road 2901. This road is not signed until you have actually made the right turn onto it.

From here on Road 2901 is signed, but it is still important to watch carefully for the number—the road takes some devious turns. About 15 miles after the turnoff from Forest Road 29 you reach Forest Road 034—which is well signed: FREMONT POINT 2 1/4. Go right 2-1/4 miles on this extremely washboarded road and park at its end, where you will find picnic tables, a vault toilet and a parking area big enough for recreational vehicles. The cabin is just a 1/4-mile walk ahead, on the left.

From Silver Lake District Office, go approximately 18 miles east on Highway 31 to just beyond Picture Rock Pass. Turn right (west) onto Forest Road 2901, and travel 18 miles to Forest Road 034. Turn left and go 2-1/4 miles to the parking area.

Travel into this remote place in winter requires experience and advance preparation. The Winter Ridge area averages three to five feet of snow, which closes Forest Road 2901. There is an alternative route—though the Forest Service advises use of a four-wheel-drive vehicle: go 400 yards east of Silver Lake Ranger Station and then south for 21 miles on plowed Forest Road 28. Here you will find an informal snow park. From here, it is an eight mile cross-country ski trip to the cabin, which will take from three to six hours depending on the weather conditions and your skill level. The trail is mostly flat, with a total rise of 1500 feet. Bring a compass and topographic map.

Elevation 7135 feet.

Map Location Fremont National Forest, Township 31 South, Range 16 East, Section 21.

What Is Provided

Two beds, two cots, woodstove, firewood, propane lantern, two-burner-table-top propane cook stove, dishes for four, lots of cooking pots, and an ax and snow shovel. No fridge.

It has a painted plywood floor, four vinyl chairs, a tiny 3x4-foot table affixed to the wall that folds down and plenty of cupboards. There is a charming old vault toilet a short distance away, and an ugly modern plastic outhouse 1/4-mile away at the parking lot.

What To Bring

Two or more propane fuel bottles and drinking water. Snowmelt can be used for your washing needs, though safe drinking water from snow cannot be assured, so have the means to treat it. Bring extra food—severe weather conditions may delay your departure.

Setting

The cabin has several screened windows that open inward, and, of course, have dizzying views. The newly added deck overlooks the glorious scenery even more daringly; its addition was a remarkable idea that will be welcomed, we are sure, by every guest and visitor.

History

In 1843, explorer John C. Fremont was the first documented traveler through this area. His journal describes his first impressions of the view from here:

> *Rising rapidly ahead to this spot we found ourselves on the verge of a vertical and rocky wall of the mountain. At our feet—more than a thousand feet below—we looked into a grass prairie country, in which a beautiful lake, some twenty miles in length, was spread along the foot of the mountain. Shivering on snow three feet deep, and stiffening in a cold north wind, we exclaimed at once that the names of Summer Lake [this area had been described by local Indians to Fremont as the land of no snow] and Winter Ridge [the area where Fremont had traveled the past few months and also the ridge where he was then*

> *standing] would be applied to these proximate places of such sudden and violent contrast.*

Fremont Point Cabin, built in the early 1930s, was once used as a summer home by fire-lookout personnel who staffed the adjacent 90-foot watchtower. The historic tower is still standing on the site, but for safety it is barricaded from use.

Around You

The view from here is extraordinary. Few structures anywhere in the country have such dazzling vistas—fewer still at $25 per night. The cabin is perched (we are tempted to say precariously, but that is not so) on top of Winter Ridge. Three thousand feet below, Summer Lake is sparkling blue, the alkali flats shimmer, and the green marsh and pasturelands soothe tired eyes. In the distance are The Steens, Hart Mountain, and Drake Peak.

"Peace and quiet with a slow breeze whispering through the pines. The best way to celebrate our six-year anniversary."

— From the Cabin's Guestbook

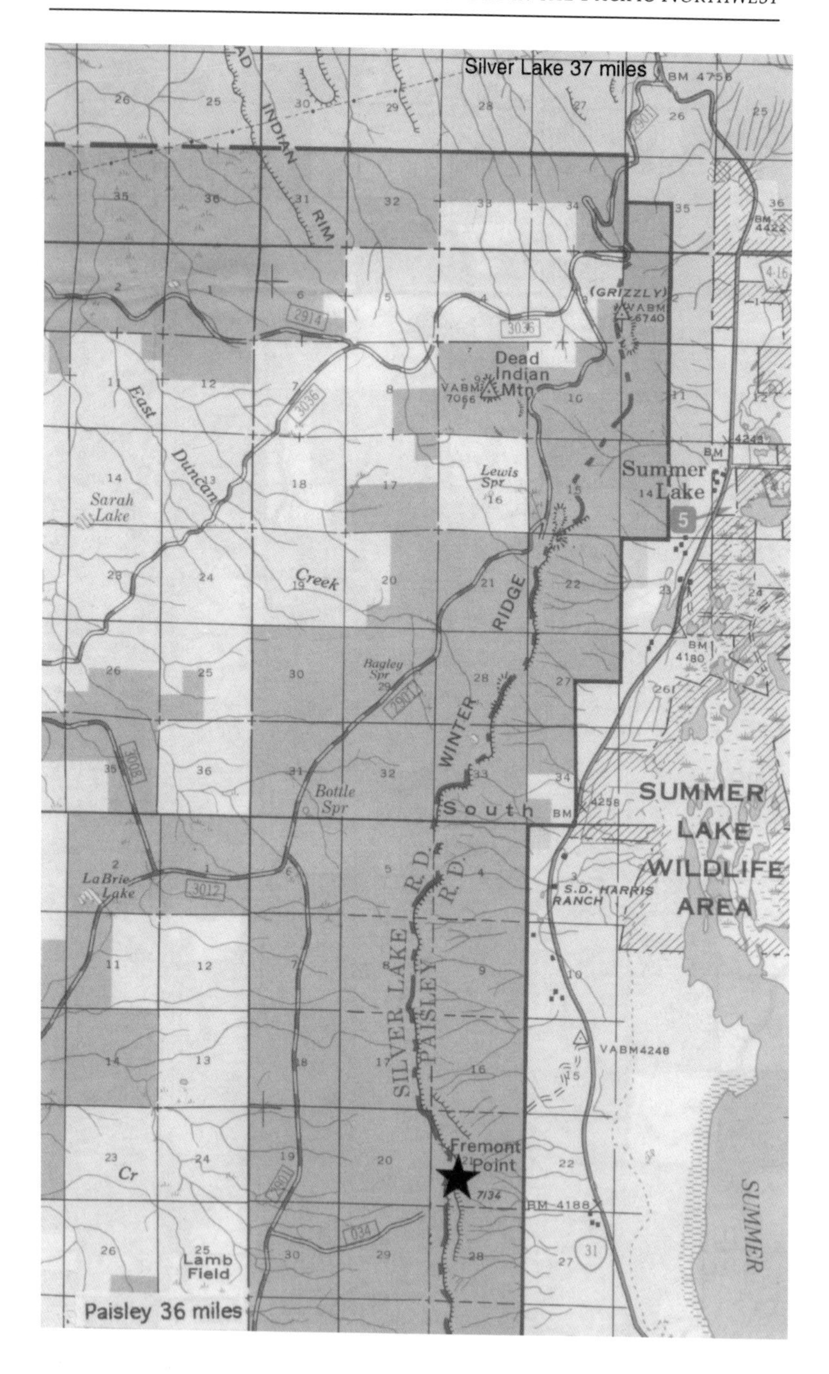

Fremont Point Cabin
Fremont National Forest, Oregon

43 Totem Bunkhouse

Emigrant Springs—Oregon State Park

"I must keep writing to remember who I am."

— FROM *Pioneer Woman*

Your Bearings

25 miles southeast of Pendleton
25 miles northwest of La Grande
65 miles south of Walla Walla
70 miles northwest of Baker City

Availability Year-round.

Capacity

A maximum of four per room. There are two rooms available, and each can be rented individually. Ideal for families.

Description

Recently built log cabin with two 15x9-foot rooms. Not secluded, but situated in a lovely park.

Cost $15 per night per room.

Reservations

Though Emigrant Springs State Park is open only mid-April to late October, the Totem Bunkhouse is open all year. For reservations, contact:

Emigrant Springs State Park
1-800-452-5687
or
541-983-2277, or 541-238-7488

How To Get There

Travel 26 miles northwest on Interstate 84 from La Grande and take Exit 234 to Emigrant Springs State Park. Or, travel 26 miles southeast on Interstate 84 from Pendleton and take Exit 234 to the park.

Map Location Emigrant Springs, Oregon State Park, Township 1 North, Range 35 East, Section 29.

What Is Provided

The bunkhouse is a log cabin, recently built, and close to the campground entrance. It is divided into two separate rooms, A and B. Each contains two bunk beds, and sleeps four. The beds take up most of the space. Outside, there are picnic tables and firerings.

The park has three 16x16-foot corrals for horses, as well as "primitive" camping for their riders. There is a horse trail that circles through the primitive area. You will also find a group picnic area and an Oregon Trail exhibit in the campground.

What To Bring

Dishes, pots, pans and eating utensils for campfire cooking; and in winter, drinking water. Snowmelt can be used for your washing needs, though safe drinking water from snow cannot be assured, so have the means to treat it.

Setting

This bunkhouse is one of two structures in this book that fulfill few of our requirements. It is neither secluded nor remote and it offers little in the way of solitude. It is, in fact, in the middle of a big campground with 18 full hookups and 33 tent sites—Emigrant Springs State Park—which is beside Interstate 84, about half way between Pendleton and La Grande. Nevertheless, it could be just the totem for weary and harassed parents who cannot travel another mile with car-bound kids who have spilled yet another can of soda—though of course it is wiser and more prudent to have made advance reservations. The good news is that the park is very accessible and is almost totally shaded all day by lovely old trees.

History

First discovered by Europeans in 1834, Emigrant Springs was a favorite camping place for emigrants on the Oregon Trail. It had fresh spring water, plenty of firewood, and lovely shade. (Regrettably the springs were destroyed in recent years by highway and pipeline construction). There is a stone monument to these travelers at the side of the highway that was dedicated by President Warren Harding in 1923. Interstate 84, broadly speaking, parallels and crisscrosses this section of the original Oregon Trail.

Around You

You are surrounded by rich Oregon history and some of the state's most beautiful scenery. To the south are the Blue Mountains, to the east the Wallowas.

If you enjoy camping at Oregon state parks, you will find yourself near many of them, including Hilgard Junction, Ukiah-Dale Forest, Catherine Creek and Minam. Day-use state parks in this vicinity are Blue Mountain Forest, Red Bridge and Battle Mountain.

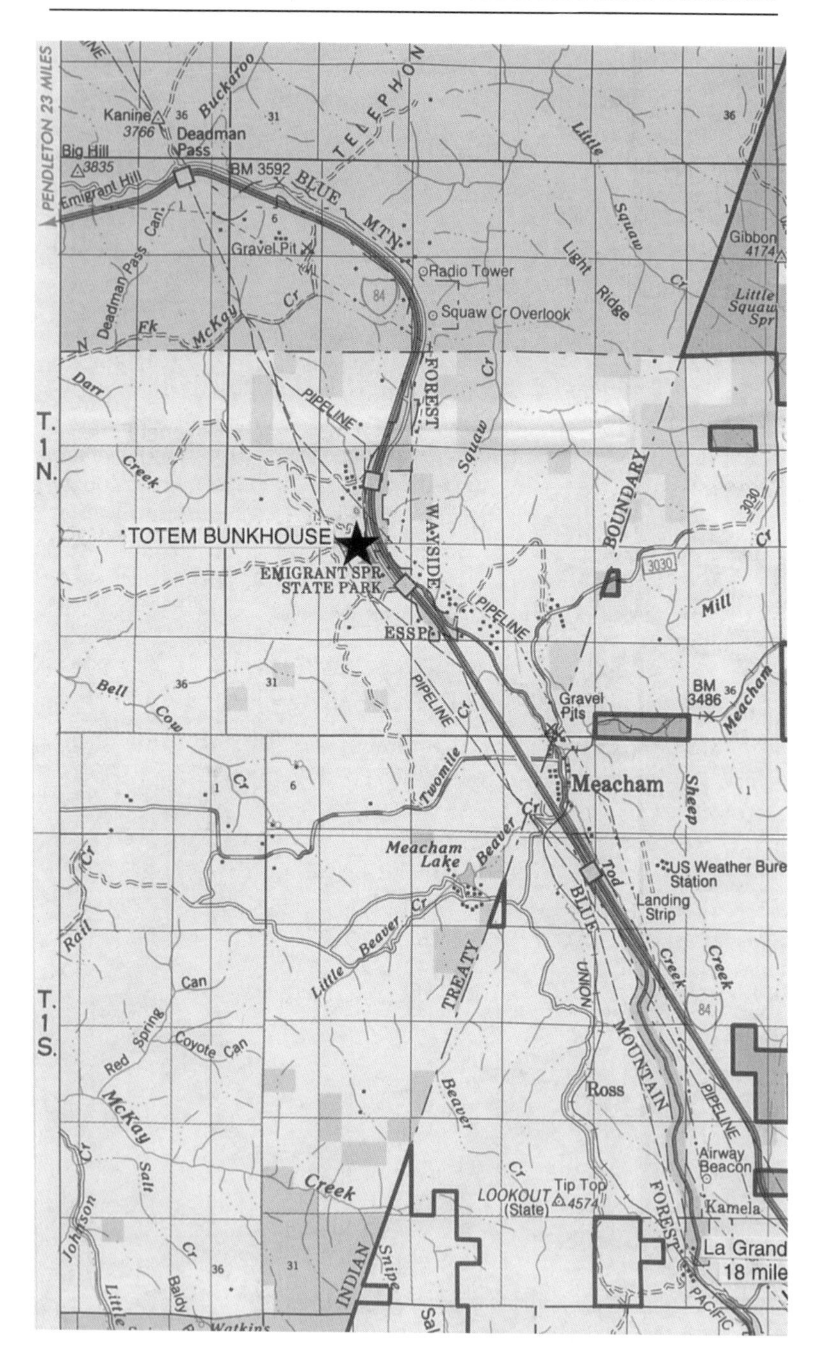

Totem Bunkhouse
Emigrant Springs—Oregon State Park, Oregon

About the Authors

Tom Foley was born and reared on a small farm in the west of Ireland, but has spent much of his life traveling and working in other parts of the globe. He now lives with his son, Nino, in Ashland, Oregon, where he works as—among other things—a father, writer, photographer, and storyteller.

Kevin Peer

Tish Steinfeld worked for eleven years as an archaeologist with the National Forest Service. She now manages her own music company, produces a nationally distributed radio program for NPR, as well as producing a regular concert series in Ashland, Oregon, where she lives with her husband, David, and her two sons, Joel and Logan.

Thomas Doty

Walking

I have met with but one or two persons in the course of my life who understood the art of Walking, that is, of taking walks—who had a genius, so to speak, for sauntering, which word is beautifully derived "from idle people who roved about the country, in the Middle Ages, and asked charity, under pretense of going a la Sainte Terre," to the Holy Land, till the children exclaimed, "There goes a Sainte-Terrer," a Saunterer, a Holy-Lander. They who never go to the Holy Land in their walks, as they pretend, are indeed mere idlers and vagabonds; but they who do go there are saunterers in the good sense, such as I mean. Some, however, will derive the word from sans terre, without land or a home, which, therefore, in the good sense, will mean, having no particular home, but equally at home everywhere. For this is the secret of successful sauntering. He who sits still in a house all the time may be the greatest vagrant of all; but the saunterer, in the good sense, is no more vagrant than the meandering river, which is all the while sedulously seeking the shortest route to the sea. But I prefer the first, which, indeed, is the most probable derivation. For every walk is a sort of crusade, preached by some Peter the Hermit in us, to go forth and reconquer this Holy Land from the hands of the Infidels.

Henry David Thoreau, *Walking*

Hiking in the backcountry entails unavoidable risk that every hiker assumes and must be aware of and respect. The fact that a trail is described in this book is not a representation that it will be safe for you. Trails vary greatly in difficulty and in the degree of conditioning and agility one needs to enjoy them safely. On some hikes routes may have changed or conditions may have deteriorated since the descriptions were written. Also trail conditions can change even from day to day, owing to weather and other factors. A trail that is safe on a dry day or for a highly conditioned, agile, properly equipped hiker may be completely unsafe for someone else or unsafe under adverse weather conditions.

You can minimize your risks on the trail by being knowledgeable, prepared and alert. There is not space in this book for a general treatise on safety in the mountains, but there are a number of good books and public courses on the subject and you should take advantage of them to increase your knowledge. Just as important, you should always be aware of your own limitations and of conditions existing when and where you are hiking. If conditions are dangerous, or if you're not prepared to deal with them safely, choose a different hike! It's better to have wasted a drive than to be the subject of a mountain rescue.

More Northwest Guidebooks

If you enjoyed ***How to Rent a Fire Lookout in the Pacific Northwest,*** and want to do more exploring in the area, be sure to look for these other Wilderness Press books:

Don't Waste Your Time™ in the North Cascades: An Opinionated Hiking Guide to Help You Get the Most From This Magnificent Wilderness by Kathy and Craig Copeland. The North Cascades offer some spectacular hiking opportunities and this book lets you experience the best that the area has to offer, so don't waste your time taking unenjoyable hikes.

Great Bike Rides in Eastern Washington and Oregon by Sally O'Neal Coates. The forgotten area east of the Cascade Mountain is ideal for bicycling. The basins and foothills offer wide-open spaces, outstanding scenery, friendly small towns, and miles on end of flat or gently rolling terrain—a great place for your next cycling vacation.

Oregon's Swimming Holes by Relan Colley. Escape from the tension and stress of the everyday world by spending some time submerged in a natural swimming hole, free-flowing lake, or stream. This guide covers swimming holes throughout the state, so wherever you are in Oregon, you're never too far from an inviting swim.

Crater Lake National Park by Jeffrey P. Schaffer. Southern Oregon's High Cascades are a glacier-sculpted, volcanic wonderland with an abundance of recreational opportunities that is bound to please almost every outdoor enthusiast.

Check your local bookstore or outdoor equipment dealer for these books, or write for our free mail order catalog:

Wilderness Press
2440 Bancroft Way
Berkeley, CA 94704
(800) 443-7227